D0463982

# OPPORTUNITIES IN
# TECHNICAL WRITING AND COMMUNICATIONS CAREERS

# Jay R. Gould
# Wayne A. Losano

Revised by
**Walter E. Kramer and Marilyn K. Kramer**

Foreword by
**Donna Roberts-Luttrell**
Society for Technical Communication

 **VGM Career Horizons**
a division of *NTC Publishing Group*
Lincolnwood, Illinois USA

**Cover Photo Credits**

*Clockwise from top left*: Cahners-Steve Hale; NTC
Publishing Group; Jeff Ellis; Foote, Cone, and Belding

**Library of Congress Cataloging-in-Publication Data**

Gould, Jay Reid.
  Opportunities in technical writing and communications careers /
Jay Gould. —Rev. / by Walter E. Kramer and Marilyn K. Kramer.
      p.    cm.—(VGM opportunities series)
  Rev. ed. of: Opportunities in technical writing today, ©1975.
  Includes bibliographical references.
  ISBN 0-8442-4128-8  (hard)      ISBN 0-8442-4129-6  (soft)
  1. Technical writing—Vocational guidance.  I. Kramer, Walter E.
II. Kramer, Marilyn K.  III. Gould, Jay Reid. Opportunities in
technical writing today.  IV. Title.  V. Title: Technical writing
and communications careers.  VI. Series.
T11.G66     1994
808'.0666—dc20                                        93-46663
                                                          CIP

Published by VGM Career Horizons, a division of NTC Publishing Group
4255 West Touhy Avenue
Lincolnwood (Chicago), Illinois 60646-1975, U.S.A.
© 1994 by NTC Publishing Group. All rights reserved.
No part of this book may be reproduced, stored in a retrieval system,
or transmitted in any form or by any means,
electronic, mechanical, photocopying, recording or otherwise,
without the prior permission of NTC Publishing Group.
Manufactured in the United States of America.

4 5 6 7 8 9 0 VP 9 8 7 6 5 4 3 2 1

# ABOUT THE AUTHORS

Jay R. Gould is Emeritus Professor of Communication at Rensselaer Polytechnic Institute, Troy, New York. He was a founder and director of the Technical Writers' Institute, established in 1953 at Rensselaer, and he is an outstanding technical writing authority and consultant.

Dr. Gould has authored and coauthored several books, including *Exposition: Technical and Popular,* a basic text in technical description, which he wrote with Dr. Sterling P. Olmsted. With Joseph N. Ulman, Jr., he is the author of *Technical Reporting,* now in its third edition, and for the American Chemical Society he has produced *Practical Technical Writing,* a cassette technical writing course. Dr. Gould has recently edited an anthology of technical and scientific articles, *Directions in Technical Writing and Communication,* and he is the series editor of *New Essays in Technical and Scientific Communication,* both books published by the Baywood Publishing Company. At present he is the editor of Baywood's *Journal of Technical Writing and Communication.* Dr. Gould is also the author of a number of magazine articles on technical communication.

Dr. Gould has served as a teacher of technical writing and as a consultant for many industrial firms in the United States and abroad. He has acted as a visiting professor in technical communication in Australia and is a fellow of the Society for Technical Communication and of the American Business Communication Association. He has received an award from the National Council of Teachers of English for his work in communication.

Wayne A Losano is Associate Professor of English at the University of Florida. He is Director of Freshman English and Coordinator of Business and Technical Communication and teaches a variety of graduate and undergraduate professional communication and composition courses.

Dr. Losano was previously head of the Department of Communication at Queensland Institute of Technology in Brisbane, Queensland, Australia. Prior to that he taught technical writing at Rensselaer Polytechnic Institute in Troy, New York, and at Lowell Technological Institute in Massachusetts. He has been a communication consultant for a number of companies, including Shell Oil Company, Marathon Oil, Xerox Corporation, and FMC Corporation.

He is a member of the Society for Technical Communication and the American Business Communication Association and has made presentations at a number of conferences. He has had articles published in the *Journal of Technical Writing and Communication, The Technical Writing Teacher,* the *ABCA Bulletin,* as well as in several film journals.

Walter and Marilyn Kramer are professional writers living in Chicago, Illinois.

# FOREWORD

We wake up to a new world every day as advances in science and technology change the way that we see and think about the universe around us. The knowledge that we as a civilization possess increases geometrically each year. As the translators between technical experts and the lay community, technical communicators are among the first to experience the newest technology, the latest scientific discovery, the medical breakthrough. These are interesting and exciting times to consider a career as a technical communicator.

The technical communicator faces the formidable challenge of quickly learning the latest innovation and then explaining it in clear, concise language and graphics to those who must use and apply this new knowledge in their daily lives. What qualifications does a career in the rapidly changing field of technical communication require?

- The technical communicator needs a curious nature and the ability to figure things out without instructions.
- The technical communicator must have a respect and appreciation for the audience with whom he or she communicates.
- Above all, the technical communicator must have excellent oral, written, and graphic communication skills. The techni-

cal communicator must be able to express the knowledge that he or she acquires in a clear and appropriate way, to make that information useful to his or her audience.

If you possess these qualities and have an avid interest in learning and communicating, the field of technical communication offers the opportunity to experience first-hand the changes of a rapidly evolving world of information.

As technical communicators, we are lifelong students of the scientific and technical world around us. We have the challenging and exciting responsibility of accurately describing and sharing knowledge. If you join us, I think you'll find it often fast-paced and challenging, but very rarely dull.

> Donna L. Roberts-Luttrell
> President, PlainTech, Inc.
> Past President, Chicago Chapter of the
> Society for Technical Communication

# CONTENTS

What is technical writing? How to become a technical writer. Examples of technical writing. Where to
find jobs. The information industry. Government-
sponsored activities. Technical journalism. Promotional writing. Audiovisual scriptwriting. Special
projects. Qualities necessary for success. Problems
of the technical writer.

Technical reports. Instruction manuals. Proposals.
Technical writing for company magazines. Technical magazines. Technical promotion. News releases. Technical advertising. Technical films and
videotapes. Technical translation. Document coordination. Teamwork in technical writing.

# CHAPTER 1

# TECHNICAL WRITING: A COMMUNICATIONS PROFESSION

## WHAT IS TECHNICAL WRITING?

The art or profession of writing goes back centuries. The original writers were poets, narrators, and historians who told all types of stories. Other than poets and historians, authors were generally classified as writers of fiction or nonfiction. These writers manipulated their words and the language to create scenes, moods, and effects so that the readers felt as though they were unobserved, passive participants to the events being described. The twentieth century saw the rise of new types of writers. Among these are technical writers or technical communicators.

Technical writers are, by necessity, anonymous authors who must remain objective and factual with the subject matter with which they are dealing. Their sole function is to deal dispassionately with facts and objects and to relate useful, relevant, reliable information to their audience. The exceptions to this rule of anonymity are people who write scientific or technical articles for newspapers, magazines, and learned publications under their own names. The language that the writers use must be simple, direct, and contain a minimum number of nonfunctional descriptive adjectives. The verbs in their sentences must be in the active rather than the passive mood to eliminate any doubt about what their writing means or implies.

Technical writing, then, is the profession of writing, editing, and preparing publications in many fields of technology, science, engineering, and medicine including articles for technical and scientific journals. The publications may be technical reports, instruction manuals, articles, papers, proposals, brochures, and booklets, and even speeches for technical meetings and conferences.

Technical writing is a useful communication tool whenever information of a technical nature must be transmitted. In addition to the USA, there are technical writers working for companies in South America, Canada, England, Europe, Israel, India, Japan, and countries all over the world. If you speak or write a language other than English, you may even be given the opportunity to work in one of these countries.

## HOW TO BECOME A TECHNICAL WRITER

With the rise of technology in the twentieth century, a new type of explanatory writing has appeared which is known as technical writing. Initially, people with scientific and technical backgrounds were given writing assignments they weren't prepared to do. As a consequence, the writing that they produced was often very poor. Not only was the work full of grammatical and spelling errors, but it was often poorly composed, stilted, and boring. It is not unusual for highly educated technicians to produce poor quality writing. On the other hand, writers with no scientific background had difficulty understanding how to present and interpret scientific data and subject matter so that the reader would be properly informed.

The need for technical writers arose because these situations almost always guaranteed poor results. Managers often did not care about the quality of the writing and were glad to give the task to anyone just to get it done. Sometimes the writing was so poor that the readers failed to derive any utilizable information from the material at hand. If the assignment was a construction manual for a process or a use manual for some piece of equipment, the results could be disastrous. After much frustration, the unhappy customers would probably give their future business to competitors.

There are two main ways to become a technical writer. The most traditional way has been on-the-job training. The ideal situation, however, in which one can learn to become a technical writer is to major in technical communications in college. In this manner, the student will not only learn to master the mechanics and techniques of writing, but will additionally be well grounded in science and technology. While not every college and university offers such a curriculum, there are a number of schools throughout the country which do, so that enrollment will not present a problem. Chapter 4 has a listing of colleges and universities where a student can major in technical writing.

## EXAMPLES OF TECHNICAL WRITING

Technical writing is becoming more varied as technology and science change and advance. As new terminology, theories, instruments, processes, and machinery come into being, others are discarded. This on-going process makes relatively new equipment and procedures obsolete almost before the packing crates are opened. Scientists in every field are constantly striving to make new discoveries. All these activities mean that technical writers must discard old paragraphs for new ones at a feverish pace. Several examples of scientific writing are cited below.

*Science News,* vol. 144, #2, 24 (1993)

When the Galileo spacecraft swung by Venus in February 1990, it was just for kicks. The encounter gave the spacecraft part of the gravitational boost it needed for its future appointment, a 1995 rendezvous with Jupiter. But the Venus flyby had an additional bonus. It marked the first time any spacecraft had imaged the cloud-shrouded surface of Venus in the infrared, recording the heat emitted by the surface.

For two 45-minute intervals near the craft's closest approach to the planet, an imaging spectrometer designed to study the atmosphere of Jupiter and its large satellites sampled the temperature on the nightside of Venus. Just as on Earth, higher elevations on Venus are colder than low-lying regions. Thus, the temperatures measured by Galileo's Near-Infrared Mapping Spectrometer (NIMS) indicate the altitude of different parts of the planet's surface.

While the Magellan spacecraft recently used radar to compile a far more detailed topographic map of Venus. Galileo's infrared instruments may provide new information to help determine the surface composition. Also, notes Kevin H. Baines of NASA's Jet Propulsion Laboratory (JPL) in Pasadena, Calif., the thermal mapping of Venus previews studies that Galileo will do when it arrives at Jupiter. The measurements also foreshadow studies with a newer generation spectrometer, now scheduled to fly on the Cassini mission to Saturn's methane-cloaked moon, Titan.

These introductory paragraphs are taken from an article about space exploration in a recent issue of *Science News*. This weekly periodical contains a variety of science-related articles describing the latest developments in medicine, pharmacology, and the natural sciences. The articles vary in length from a single paragraph to several pages. The subscribers are primarily scientists who want a quick overview of what is occurring in other sciences. However, many nonscientists are also regular readers since the articles are well written and understandable to the layman.

*Chemical and Engineering News,* vol. 71, #343

### Synthesized Carbon Nitride Film

Chemists at Harvard University have prepared carbon-nitrogen thin-films that appear to consist primarily of beta carbon nitride, a material that theory predicts could be harder than diamond. The synthesis represents a significant achievement in basic materials research and could have important engineering applications.

"If beta carbon nitride is harder than diamond, it could have a wide range of uses," says Charles M. Lieber, the Harvard chemistry professor who directed the research. "Making this compound also was a test of our synthetic skill. It is very challenging to prepare what is very likely a metastable material from components—in this case graphite and nitrogen—that are themselves very stable."

Lieber points out that nitride compounds are, in general, difficult to produce because of the stability of molecular nitrogen. "You cannot simply heat graphite and nitrogen together and expect to get carbon nitride," he says. "We had to design a strategy to generate species and trap them kinetically. Organic chemists routinely use kinetic versus thermodynamic control to carry out elegant syntheses of complex molecules, but solid-state chemists have relied almost entirely on thermodynamic control."

This article describing the synthesis, properties, and potential uses of carbon nitride is an example of a highly technical article. The writer is giving an account of a breakthrough in solid state chemistry. Carbon nitride, whose properties are widely known, is a rare compound because it is so difficult to synthesize. The author must be able to handle highly complex ideas and terms and still be able to write an interesting story which is easy to understand. He does have one advantage: most of his readers are chemists. Nevertheless, few of them will have detailed knowledge about the field of work described.

*Chicago Tribune, Discovery Section,* July 18, 1993

### Fossilized Dinosaur Egg Gets Some X-Ray Attention

Some of America's most sophisticated X-ray technology is being trained upon a fossilized dinosaur egg scientists say is about 65 million years old.

The egg was dug up in Southern China and is now at the Lawrence Livermore National Laboratory in California where scientists are using computerized tomographic X-ray equipment to make 3-D images.

The California researchers, who are working in conjunction with Chinese scholars and those at the University of Notre Dame in Indiana, also are applying high-energy X-rays to make two-dimensional images of the egg, which is cracked at one end and contains an embryo.

J. Keith Rigby Jr., a Notre Dame paleontologist, said that because the fossil comes from an area of China where few dinosaur remains have been found, the scientists hope this find may cast new light on a type of dinosaur they know very little about.

This article was written by two science writers for the *Chicago Tribune* who regularly prepare a variety of feature articles for the paper about scientific topics of current interest. They are expected to write articles for the public explaining the latest developments in the particular field of science they have selected for their subject. They must be very careful not to appear condescending to their lay readers who often have little or no scientific knowledge. Yet, the readers must be able to understand what they read. Many major newspapers such as the *New York Times, Washington Post,* and *Los Angeles Times* pride themselves in having similar staff writers.

*Sargent-Welch '93–'95 Catalog,* page 586

**Suspension Galvanometer**
**2690A. Galvanometer, Student, Mirror Type**

Provides experience in manipulating and reading a mirror and scale system. With scale removed, it may be used in demonstrations to indicate current by deflection of a light beam. Sensitivity is about one microamp/div.; coil resistance, 80 ohms. Coil and attached plane mirror are suspended by phosphor-bronze ribbon between poles of a strong permanent magnet. Zero and coil elevation adjustments are at top coil clamp and binding posts at rear. Glass front of case is removable for access to all parts. Leveling screws are located in two legs of tripod; third leg is longer and supports scale and eye aperture. Silvered area on glass plate near mirror superimposes a fixed vertical reference line of image of scale, providing an index for reading the scale. Height, 27 cm. With instructions.

Here we have a paragraph taken from a scientific catalog. The catalog is intended to take the place of a salesperson by giving a prospective buyer a brief, concise, and accurate description of a scientific apparatus under consideration for purchase. The catalog must give potential customers complete information on all its items so that they will be able to order precisely what they want. The emphasis in this type of writing is to provide necessary information using a minimum number of words. For this reason, complete sentences are not always used.

Four types of technical writing have been presented to give you an idea of its scope. As you can see, the field is very broad and may cover everything from the preparation of an article in a specialized journal to a syndicated story in a large, influential newspaper. Technical writers are expected to be jacks-of-all-trades. An examination of the job descriptions given in the job ads in the next section shows that they are expected to perform a whole host of tasks from writing complex papers to mundane notices for company bulletin boards.

## WHERE TO FIND JOBS

Everyone wants a job for which they have trained and educated themselves. The question is where do you find them and how do you go about getting one. Every college and university maintains an employment office. In the past, these offices would only lend assistance to their own students and alumni. Today, many of them extend this courtesy to any qualified college graduate, alumnus or not. Another potential job referral source is the Society for Technical Communication in Arlington, Virginia. A letter or a telephone call will allow you to get what information they have posted. Also, they will refer you to the local section office nearest you which may have job openings not listed with the national headquarters.

One of the richest sources for available jobs is the Sunday help wanted section of major metropolitan newspapers. The papers carry a variety of advertisements for job openings. The following examples, taken from a Sunday edition of the *Chicago Tribune,* are typical for such newspapers.

### STAFF EDITOR

The College of American Pathologists, an internationally known medical society, is seeking a Staff Editor to join our editorial staff at our Northfield, Illinois headquarters. Candidate will utilize desktop publishing techniques, and be responsible for editing, design and producing a variety of in-house and nationally distributed publications, including: scientific and socioeconomic trade publications; brochures, manuals and pamphlets; questionnaires and advertisements; etc. Mac experience helpful. The ideal candidate will possess a B.A. in liberal arts or a related area, and 1 plus years of related editorial experience. Health care or association experience would be helpful.

### TECHNICAL WRITER

Leading consumer electronics manufacturer in far western suburbs has an opening for an experienced Technical Writer. Duties include research and accumulation of data for the preparation of technical manuals describing sequential procedures for the use of our products. Will also assist in the production of public relations material, newsletters, and public service pamphlets. This position requires 1 to 2 plus years of technical writing experience, knowledge of data processing concepts, and a college degree in English or journalism or equivalent experience. Proficiency with word processing and desktop publishing software is essential. Strong written and oral communication skills are expected. Experience with Microsoft Word and Pagemaker a definite plus.

**TECHNICAL WRITER**
Large Loop law firm has opening for a technical writer. Background in computer-related writing, including courseware and user documentation required. Must have analytical skills and an end-user orientation and be comfortable working in a team environment. Bachelor's degree preferred. We offer competitive salary and benefits package.

**TECHNICAL WRITER**
LPC, a Pitney Bowes Company, is a leading developer of IBM cross-platform software. We design systems that support the market planning and high volume mailing applications of Fortune 1000 companies. Our corporate headquarters in Lisle has an excellent opportunity for an experienced writer with a strong technical background. You will write and edit technical documentation for our existing software products, including user guides, training manuals. Your responsibilities will involve interfacing with technical and customer support personnel to ensure documentation is user friendly as well as technically accurate. The position requires 1 to 2 plus years of technical writing experience, knowledge of data processing concepts and a college degree in English or journalism or equivalent experience. Proficiency with word processing and desktop publishing software is essential. Strong written and oral communication skills are expected. Experience with Microsoft Word and Pagemaker a definite plus.

These four employment ads clearly illustrate the great diversity in the term "technical writer" and what companies expect of people whom they hire. Certainly from the many tasks assigned to technical writers by their employers, the positions are multifaceted and anything but boring. A law firm that hires a technical writer will have different expectations for their employee than an engineering company. So, it is best that you understand that the type of person a company hires will depend upon the nature of its activities. Also, note that all these ads require computer-literacy, another important consideration in preparing for a technical writing career.

## THE INFORMATION INDUSTRY

The need to transform masses of data into structured, intelligible information has created an ever growing industry called the information industry. Actually, this industry has no identifiable plants, buildings, offices, or factories as such. Rather, it exists in a myriad of companies, governmental agencies, laboratories, colleges, and universities in the form of individuals and departments whose sole function is to produce objective, understandable information for laymen, regardless of the subject at hand. Consequently, technical writing is not limited to topics connected with science, technology, and engineering. The ability of technical writers to convert raw data into utilizable information has received widespread recognition as a special skill and a talent. Companies which must sell the products they manufacture now realize the importance of having technical writers on their staff. Service organizations likewise must inform the public about their activities in order to sell what they have to offer.

A table of contents of a typical industrial manual which accompanies every manufactured item usually includes: a general description, a theory of operation, instruction on installation, adjustment, and operation, and finally several drawings of the item. An instruction manual must be prepared to tell the customer how to install and operate the purchase safely. Often troubleshooting suggestions are included. The people who prepare these manuals are among the most skilled in the technical writing profession, for they must be thoroughly familiar with the equipment and they must write in a clear, concise, simple, and unequivocal manner. The installers and operators are oftentimes people with a limited education and vocabulary. It is the writer's responsibility to oversee the manual from inception to final printing. Approximately one-third of all those who are employed as technical writers work exclusively on instruction and maintenance manuals.

## Proposals

The writing of proposals is very important to most manufacturing companies. A contract usually precedes the start of any manufacturing operation. The contract may be between the company and a government agency, between the manufacturer and a supplier of parts and materials, or between the manufacturer and the company which is buying the finished product. A contract proposal is prepared in which the product and the standards to which it is supposed to adhere is submitted to the purchaser. The two parties to the contract then work out the final purchase terms. Contract proposals probably rank next in number to reports and manuals.

## GOVERNMENT-SPONSORED ACTIVITIES

The Defense Department maintains a Defense Documentation Center (DDC) in Alexandria, Virginia which keeps track of all the scientific and technical information coming from the industrial and research activities it sponsors. Many universities and private research foundations also have government-sponsored programs. Additionally, the DDC files reports on all writings which it deems significant, even on work not government-funded. This is a staggering task. In this country alone, there are thousands of industrial concerns ranging in size from AT&T to small shops that manufacture just one or two specialty items.

Many of the larger companies which manufacture highly complex machines such as combat aircraft, computers, and nuclear power plants for submarines also maintain huge research laboratories. Written reports on all of the activities including ideas for new or modified products must be submitted. The organizations which submit these reports do so not only because they are required but also because they hope that this information will lead to new contracts and revenues.

The flood of information is so great that the DDC has to put all this onto computer discs and tapes to conserve space. In the majority of these situations, technical writers are employed to prepare the reports from the data and information provided by an engineer, chemist, or physicist. It is then their task to organize, prepare, and distribute the final reports.

## TECHNICAL JOURNALISM

We are literally surrounded and inundated by a flood of literature called technical journalism. The number of publications, periodicals, magazines, trade journals, and newspapers is so great that we take their existence for granted and hardly notice their presence. There are so many scientific and trade publications, in fact, that it is impossible to cite a number with which everyone will agree. In addition, new publications are constantly appearing.

The text and subject matter range in complexity from the highly scientific *Journal of Organic Chemistry* to the very simplistic magazine *Popular Mechanics*. The *Journal of Organic Chemistry* is found only on the book shelves of chemistry and technical libraries and in the offices of chemists. Naturally, most of its readers are organic chemists. It is not generally available to the public at large. Similar journals would be the *Journal of the American Chemical Society,* the *Journal of the American Medical Association,* the *New England Journal of Medicine, Chemical Engineering, Physical Reviews,* among many others.

In scientific journals, the writers are generally the scientists who are involved with the work. Their names appear directly under the titles. In academic institutions, papers presented for publication are usually written by the graduate students who actually did the work. The professors in charge seldom soil their hands with writing papers. Sometimes, graduate students who had nothing to do with the work do the actual writing and remain anonymous. In industrial concerns, the preparation of papers is usually done by technical writers to enable the authors to engage in "more productive work."

Papers presented to scientific journals for publication undergo a process called peer review. After the paper has been presented to a journal, copies are sent to recognized authorities in the field. They carefully examine the paper for originality of work, for correctness of experimental procedures, and for the validity of the claims. If they are not satisfied with the paper, they reject it and state the reasons for their rejection. They tell the authors what they must do to satisfy their objections. As a rule, the authors will comply with the objections and do additional work to make the paper acceptable for publication. It is not unusual for a paper to be presented and rejected several times before it is finally published. This process reduces plagiarism, fraud, and insures scientific integrity.

There are many scientific and technical magazines sold on newsstands and kiosks to the general public. Among these are *Scientific American, Science, Omni, Psychology Today, Popular Science,* and *Popular Mechanics.* The first two magazines often contain articles every bit as sophisticated and mathematical in content as those published by scientific and medical associations. Magazines like *Popular Science* and *Popular Mechanics* are intended for people who know little about science or technology, but who are, nevertheless, interested in these matters. The writers for these magazines must be able to interpret and present scientific material in such a way as to make it understandable and attractive to their lay clientele. The success and longevity of these magazines speak for themselves.

Included in the category of technical journalism are advertising brochures, pamphlets, and catalogs published and distributed by companies to sell their products and to attract new customers. Doctors are regularly visited by pharmaceutical salesmen who give them this literature plus free samples of the latest approved drugs. In a similar fashion, agricultural companies descend upon farmers with salesmen bearing advertising material to induce them to use their seeds, fertilizers, pesticides, herbicides, animal feeds, and farm implements. Although this promotional material does contain a lot of factual, reliable data and information, it must, nevertheless, be classified as propaganda.

Finally, we come to the realm of trade journals and house organs. Almost every technical, commercial, trade organization, and business association, no matter how small, publishes a monthly magazine which contains material of interest to its members. The articles therein may discuss such subjects as the effects of crime, tax laws, or new marketing techniques on business.

Every company that has publicly-traded stock issues an annual report to its stockholders. Some of these reports are magnificent beyond imagination and are often twenty pages long. A lot of work, money, and effort are expended in preparing these annual reports. The reports are not only intended to inform but to impress the reader. In addition to these annual reports, many large companies publish and distribute house organs to their stockholders and libraries. An example of this is the *Lamp* published by the Exxon Corporation. This magazine in a narrative form presents articles about its far-flung operations, new technology it is adopting, its research projects, and its *pro bono* activities. These magazines are printed to present a favorable image of the company.

## PROMOTIONAL WRITING

Many technical writers prepare sales or promotional literature for a large variety of merchandise such as automobiles, home appliances, and consumer electronics goods. Brochures are printed by the manufacturer for potential customers containing lots of technical data about the product they are considering buying. The brochure will also contain other information describing its virtues and attractiveness. This is best illustrated by what happens when you visit an automobile showroom to buy a new car. Not only does the salesman describe the merits of the car you are interested in, but he also hands you a very attractive brochure describing its engineering, performance, safety, and appearance features. He then points out certain items in the brochure he wishes to stress, hoping to make the car purchase irresistible.

Technical sales literature comes in many forms. One example is cited above about sales brochures for cars. Another could be a pamphlet describing the construction of a turbine for a government project. Or it could be information for a new product soon to be released to the industrial market. The publicity department of a company may handle all these activities—preparing news releases for trade journals, newspapers, and magazines, and brochures to be sent to potential customers. These written items usually combine sales appeal and technical information.

As a member of the public relations department of a company, you may be asked to determine what your customers think of your company's products. Then you, along with others in your department, can expect to be given the assignment of planning a sales promotion campaign not only to maintain the loyalty of your customers but also to win over the affections and dollars of your competitors' clients. Here, for example, are some representative projects in which you may participate.

- Developing brochures, press release materials, and other pieces that explain technical products and processes.
- Preparing feature articles for trade and technical magazines that describe new technology.
- Writing speeches and presentations that will be delivered by organization executives at various meetings.
- Preparing the company's annual report.
- Developing audiovisual presentations.
- Preparing position and technical papers for presentation to governmental agencies.

Today's corporate managers and executives seldom have time to get involved in the actual nitty-gritty of speech writing and preparation details. They will decide on the theme and the main topics to be stressed. Also, they may offer guidelines as to how they wish to present and embellish their speech. The rest is up to the publicity department to "give the boss what he wants." If the speech is deemed important enough, outside resources and assistance may be enlisted for the preparation.

A technical advertising copywriter has a job category closely related to that of a promotional and publicity writer. The technical writer who works in this phase of the profession is usually employed by an independent advertising agency whose clients are companies that need outside help in advertising their products.

## AUDIOVISUAL SCRIPTWRITING

Who doesn't have visions of being a screen star or a soap queen on a popular TV serial? Alas, in your capacity as a script writer, you'll be behind the camera, way behind. However, you will lose your anonymity. It is customary to give every person who is a major contributor to the film or tape a credit. The preparation of training, instructional, and documentary films and tapes is a huge business. "Who," you may ask, "produces and uses these?" Generally, they are produced by private film and television studios, or even by large companies which have their own production facilities. The films and tapes are broadcast by private or public television stations. Some films, such as the Civil War documentary by Ken Burns, are so superlative that they attain widespread fame.

Audiovisual productions cover a wide range of topics. Uses of these films include training personnel in industry, hospitals, and the armed forces. Private companies produce films which they distribute to their customers to teach their personnel how to install and operate newly-purchased equipment. Schools will use tapes and films for instructional purposes. The types of instructional films shown will obviously vary with the age of the students, the school they attend, and the course of instruction. Audiovisual presentations are popular at all educational levels from elementary through graduate school and for adult education.

Here is what you can expect when producing a training video becomes your next project. Videos are one of the most popular methods of training. They are effective when the script is well written and holds the viewer's attention. The video must complement the printed material so that it becomes a visual aid to encourage the viewers to think about what they have just seen. Keep in mind that the information presented in the video must reinforce the concepts contained therein. Finally, consider how the visual presentation will provide the basis for class discussions about its subject matter. This is the purpose of audiovisual projects.

In preparing a script, think about how it will look to the viewer and what effect your material will have on the audience. Once the script is written, consider what cinematography techniques you could employ to enhance the value and effectiveness of your work. This is an opportunity to give your creative imagination a workout. After all, true cinematic genius does not live in Hollywood alone. To create a great script, consider the following suggestions:

1. Determine the nature of your audience.
2. Do extensive research on your subject before you write a single line.
3. Simplify your message.
4. Tell a creative, visual story.
5. Write narration with eloquence and dialogue with character.
6. Work with the director and the production crew throughout the entire filmmaking process.

A smart scriptwriter will take advantage of the expertise of everyone associated with the project. By soliciting their advice, cooperation, and suggestions, you will make your job easier and insure the success of your film. A successful video will make the audience receptive to your message and leave them feeling that they have learned something valuable. In conclusion, the following want ads show that interesting opportunities await those who have the requisite skills.

### VIDEO SCRIPTWRITER
Chicago film and video company needs well-traveled free lance writers to write travel documentaries. Must have a minimum of 5 to 10 years of film or video writing experience. Competitive advances and royalties guaranteed to those contracted.

### ADVERTISING COPYWRITER
Interested in fine and performing arts, foreign films, documentaries, and business training programs? Chicago video distributor seeks versatile and self-disciplined copywriter with college degree and 3 to 4 years experience to write copy and assist with video scripts and preparation. Applicant must have excellent verbal and written English skills and be detail oriented.

## SPECIAL PROJECTS

Industry has found a need for special project writers. The vast field of research depends upon the interchange of ideas. Because of this, many companies encourage their engineers and scientists to prepare papers for presentation at conferences and seminars. The preparation of a paper is a time-consuming process.

Some companies, particularly those with large research staffs, have a technical publications department whose function it is to assist their staff members to prepare and present papers. The writers may use report drafts to write the whole paper, subject to editing by the presenter. Graphs, photographs, projectuals, and slides are often an integral part of the presentation. In addition, the writer may coach the presenter in delivering the talk if he is shy or a novice at this activity. Many executives now employ technical writers as special assistants. These assistants, combining a flair for writing with a sound scientific background, help top management officials by writing progress reports for them or reviewing their speeches.

One of the biggest jobs in technical writing results when the writer becomes a supervisor or a department manager. Managers may supervise not only technical writers, but also the people engaged in illustration, graphics, photography, and distribution. In short, supervisors mobilize their departments to produce the specified printed matter, be it a report, manual, or technical article.

With the continuous introduction of new technology, the communications field is an ever-changing and expanding field. Technical writers will be preparing lectures and instructions to be delivered from video cassettes, developing programs for data storage and retrieval systems, and performing a multitude of other communications jobs too numerous to mention. Traditional pen and paper reports have all but disappeared. The technical writer now uses word processors. Even the time-honored typewriter has gone to its final resting place. The writer must be prepared to adopt and adapt new communications and computer techniques as they present themselves. The technical writer, like the rest of us, must change with the times. He or she must also remember that whatever writing and communication tools he or she presently has are merely transient, waiting to be supplanted by something newer and perhaps better.

## QUALITIES NECESSARY FOR SUCCESS

Everyone who wishes to make a career for himself or herself in technical writing must ask, "What kind of person should I be to succeed in this field?" and "What kind of personality traits should I have?" Some of these are obvious and are similar to traits that make people successful in any business field or profession. You should be persistent and forceful, but not obnoxious, in seeking the information you need. It goes without saying that you should enjoy writing. You must be a self-starter with a keen analytical mind who is at ease with management and can speak their language. Finally, you must have the capacity to assume responsibility and be willing to learn continuously about your field. Refusal to stay on top of your job is the quickest route to the unemployment office.

The success of an engineering enterprise depends on the cooperation and interaction of administrators, engineers, and technical communicators. An engineering firm's administration must consider the individual personalities and the interaction of all the people it has on board. They should look for prospective employees with the following characteristics:

1. All members must regard themselves as being players on a team, with each one having a specific function. A large part of each participant's time may be spent outside their field getting information and data from engineers and working with other publications personnel, printers, and illustrators.

2. They must be tactful. The job of editing requires a high degree of diplomacy. The less people write and the less skillful they are, the more sensitive they will be at taking criticism about their literary craftsmanship. For generations, engineers have been told they do not write well. Consequently, they may resent being criticized by professional writers. Tactfulness does not imply cowardice. It simply means that the writer has to cultivate a rapport with engineers and scientists, and know how to offer constructive corrections and suggestions about their writing.

3. They must also be capable of dealing with details and minutiae. Many times the technical writer is anxious to get the job done as quickly as possible, but quotations must still be authorized, statistics checked, and all kinds of calcula-

tions verified. Very often a highly technical scientific project will require gaining an intimate understanding of the subject before proceeding with the actual writing. The report that is subsequently written may require collecting and compiling large amounts of technically accurate, detailed data prior to its publication. If you are averse to working through the unglamorous aspects of writing, you probably will not make a good technical writer.

Another important aspect of technical writing is the constant necessity for self-improvement. In short, the most successful person is an eternal student. In this world there is always competition for your job and means of livelihood. There is always someone not only within your department but elsewhere in the company waiting and wanting to take your place.

There is no way around it. Technological obsolescence is occurring at an ever-increasing pace. Machinery, equipment, knowledge, and skills become obsolete. The only way you have of combatting this situation is to set up a program of constant self-improvement for yourself and stick to it. The following are some of the ways you can accomplish this.

- Enroll at a local college or university night school for an advanced degree.
- Enroll at a local college or university night school and take some relevant courses each semester.
- Keep up with the literature in your field.
- Enroll in company-sponsored courses (if offered).
- Attend as many technical seminars and conferences and take as many short courses as you can.

Technical writers have a very important function: namely, to teach others about their profession. One technical writing graduate made the following statement about her job: "A very interesting thing has happened to me. I'm giving a thirteen-week course in technical writing for my department, the civilian Atomic

Power Department. I also edit reports and classify and abstract patents on nuclear-powered reactors."

It is likely that this technical writer will go far because she has jumped into a new area and has demonstrated her ability to explain, describe, and illustrate. These characteristics will help her immeasurably in preparing training programs.

You must also develop a sense of objectivity and you should be able to place things in their proper perspective, unaffected by personal prejudices. The young man or woman who starts out in a new position saying, "This is not the way we did it at another company" (or at school), is in for a rude awakening. Industrial publications must be processed in the shortest possible time. The publications department may have already worked out a procedure that fits the company perfectly. The new technical writer should be able to recognize this and adjust to it.

You must like the basic tools of your trade, meaning, that you like to write. This is not a new idea by any means. But, some people overlook it when choosing a career. The greater your mastery of words, punctuation, and grammar, the greater will be your skill in a variety of writing modes. It will then follow that your chances for success will also be greater. It has been shown time and again that those who are articulate both in oral and written presentations are better equipped for good jobs and for regular promotions than those who do not communicate well. Every report we have seen, every questionnaire filled out by technical communicators now holding important management positions, attests to this fact: the author knew how to write well, how to present proposals, and how to speak convincingly.

We could go on almost indefinitely citing useful characteristics of the technical writer. But if there is one characteristic that seems to predominate, it is an interest in both the arts and the sciences. In dealing with graphic artists and illustrators, you must have some appreciation of their skill and ability. You should know the basic principles of good composition in an illustration or a photograph and why certain kinds of graphics are appropriate for one situation but not for another.

The recent introduction of the computer and all of its accessories into the workplace has raised the expectations of what is required and demanded of novice and veteran technical writers. At a recent executives' meeting in a large company which hires hundreds of technical writers, a lengthy discussion ensued concerning what the company expects of its writers. The list of requirements and expectations was staggering. In addition to writing and editing skills, the company expects that its writers will become familiar with graphics management, especially the layout and design of documents. They must be able to turn out documents that the reading audience for whom they are intended will find acceptable. The company expects that its technical writers will be totally computer literate and use the latest hardware and software in performing their tasks.

## PROBLEMS OF THE TECHNICAL WRITER

The phrase "media intake" implies that communication has at its disposal more media and channels of information than were conceived possible just a few short years ago. As with the scientist, more information reaches and is available to the writer than can be assimilated in a short period of time. This poses great frustration. The writer literally feels inundated by this ever increasing tidal wave of information. In this vein, the late President Kennedy is reported to have said, "I'm reading more and more and enjoying it less and less." Thus far, no simple solution or rational answer has been found for this problem. Earlier in this chapter there was a statement which said, "A technical writer must become an eternal student." You will have to spend many hours reviewing mountains of information just to keep up-to-date. There is no such thing as resting on one's oars. There is the story of the Hollywood agent who was fired by an actor for failing to get him a choice role. The agent protested and proceeded to give the actor a long list of accomplishments on his behalf. The actor responded thusly, "Yes, I know. But, what have you done for me lately?" Unfortunately, this little vignette, funny as it may sound, is true everywhere people work for others. So, to insure your continued value and employability, you must stay current and on top of things in this fast-moving world. People can be replaced in the work place very rapidly.

You will find that in a technical writing career you are dealing heavily in human communications. Therefore, you must maintain effective personal interaction. Often you may wonder why you aren't getting responses to your requests, why you are not receiving the same information other members of your staff are getting, or why some of your meanings are being distorted by your readers. In order to minimize these problems, you must keep the lines of communication between those with whom you interact open, clear, and active.

Other communication problems may arise between you and the people with whom you associate on your job or those for whom you write. It is not at all unusual for scientists and engineers to think and behave differently than you do. The same words which both of you use may have completely different meanings to each party. This may also apply to your readers. You will be expected to adapt yourself to them since most of them won't change their ways for you. Marshall Field, the great entrepreneur, once said, "The customer is always right. So, give the lady what she wants." Because of this, you must learn how to address your words and writing and to use a language appropriate to your audience. All these problems emphasize that the field of technical writing is an ever-evolving profession with ever-changing demands.

# CHAPTER 2

# THE JOB OF THE TECHNICAL WRITER

At a meeting of the Society for Technical Communication, John L. Simons, a technical editor at Catalytic, Inc., began a speech this way:

> The technical editor makes decisions constantly. Every time he reviews a draft, he must decide whether the words and their arrangement meet the quality criteria of his organization. He must decide whether to accept the material, reject it, or change it.
>
> The pressure of deadlines brings an added dimension to the rejection vs. change decision. If the final draft has defects, the editor's decisions become critical.
>
> The technical editor makes decisions all day long, every day. Admittedly, these decisions may be minor or routine, but there certainly are many of them. Strictly speaking, these decisions concern the form of a document rather than its substance. The technical editor is called on to decide whether the words in the draft versions are the correct words—decide whether the words are spelled correctly, whether they are the best words for conveying the meaning.

Job descriptions sent out by two companies looking for technical writers will tell you better than anything what is required for success in this field. The following description is from an institute which serves a number of companies in the construction business.

### Job Duties and Responsibilities

- Secure and write construction-related articles through visits to design offices, building sites, and individuals involved.
- Report construction news including that obtained at sponsored conventions.
- Rewrite and amplify press release material received from outside sources.
- Edit construction-related papers and articles obtained from outside sources.
- Applicants should have experience or training in technical writing as well as in engineering and science.

To show the widespread opportunities in technical and professional writing, we are including a job description from a pharmaceutical company:

A leader in manufacturing and marketing pharmaceutical products on a worldwide basis, we seek a scientifically-oriented writing professional who can initiate, develop, and coordinate our various communication requirements as follows:

- Prepare scientific reports from raw medical data, investigators' brochures, package inserts, medical abstracts, and product information summaries.
- Review manuscripts for publication and marketing pieces for technical accuracy.

The successful candidate should have a B.S. in chemistry, biology, or pharmacy, plus three years' writing experience.

One view of the versatility of the technical writer's job was given by Jay J. Goldberg, a technical writer at Hoffman Electronic Corporation and a frequent writer on communication.

Large companies produce specialists. One person writes theory, another does parts lists, another coordinates with typists or illustrators. Most people are unaware of tasks outside their small area of activity.

In a smaller company—and this, it seems, is where many of the future jobs will be—a publications person, particularly a writer, must be a generalist. In a smaller publications department, say from five to fifty people, a writer is usually responsible for his or her project from scheduling through printing. He or she must outline, write, lay out rough schematics and drawings, coordinate with clerical help, direct the photography of equipment, make parts lists, plan the final layout of the book, and prepare the printer's assembly sheets.

In highlighting these two career paths, Mr. Goldberg poses an interesting question: Is it better for you as a future writer to train for specialized jobs, or should you become more versatile and general? We will be examining this question later in this book.

Gibson A. Cederborg is a technical editor at the Naval Explosive Disposal Ordinance Facility at Indian Head, Maryland. Based on his experience and with some humor added, Mr. Cederborg's words make a fitting summary for this section:

> We can say that the qualified technical editor is a sort of specialized jack-of-all-trades. He or she melds literature and science, understands people, implements management policy, is objective, steers steadfastly toward a goal, and remembers minutiae. He or she can adroitly answer the writer's questions such as:
>
> Why are you taking the zap out of my draft?
>
> What are you asking me, "What are the test objectives for?" when I've already explained them in my report?
>
> What is your reason for requesting a rewrite with different conclusions?
>
> Why are you deleting the entire paragraph on potted electronic circuits?

At this point, before we go into specific types of technical writing, we should make one point very clear. The industrial writer over the past few years has been at the center of advances in communication. When you join an organization these days, you find yourself involved in new methods of using graphics, new methods of transferring both ideas and facts to paper through word processing and computerized devices, and new methods of translating from one language to another.

The technical writer becomes a member of an organizational team. Frequently he or she knows, through sensitivity to communication, ideas that are only in embryo form in the company long before the engineer is aware of them. Through contacts with marketing people, the writer finds out what research is being done and plans for new products. Despite the long-range plans, the technical writer must handle the everyday, bread-and-butter types of communication, which we will discuss.

## TECHNICAL REPORTS

In any library, perhaps in a university or government installation, you will find thousands of technical reports. The number of such reports on file throughout the world is mind-shattering, so astounding that methods are constantly being sought to reduce the space that all of this written material requires.

At the same time that reports represent a very large part of the publication output of a company, they also represent a large part of the time taken up by many technical writers and editors.

This is the cover of a typical research report:

NASA CONVENTION REPORT
DIFFERENTIAL THRESHOLDS
FOR MOTION IN THE PERIPHERY
by
James M. Link and Leroy L. Vallerie
Prepared by
DUNLOP AND ASSOCIATES, INC.
for Electronics Research Center
NATIONAL AERONAUTICS AND
SPACE ADMINISTATION

As you look this over, you will notice how complex this report seems to be, particularly regarding the people and organizations involved. Without knowing exactly what its history is, we can speculate how the report came to be written. Perhaps as the result of a proposal, NASA became interested in the subject stated in the title. A contract was then drawn up with a research group called the Electronics Research Center. Two men, Drs. Link and Vallerie, then started work on the experimentation. At some point, the job of reporting the research was given to a service company, Dunlop and Associates. They eventually produced the report to be handed over to NASA. The unseen hand, or perhaps we should say pencil, behind the writing of this report was undoubtedly a technical writer.

A report of this kind is usually preceded by a series of progress reports, short documents indicating what has been accomplished at stated intervals and submitted to the contracting agency. Sometimes a technical writer helps with this work; at other times it is done entirely by the researchers. If the project is complicated and takes a long time to complete, a periodic report is written; this report consolidates a number of individual progress reports. The periodic report may also be edited by a technical writer. After the material submitted by the scientists is put together, the technical editor will have a hand in it. In the case of the NASA report, NASA spells out certain specifications for the format of the report. This format must be checked by the editor, along with grammar, punctuation, and other stylistic features.

Report writing is one of the principle categories of technical writing. As technology increases, as companies become larger and larger, so do the written records and communications. Only by means of reports can many of the larger companies coordinate their various activities, especially if research is a principle occupation. The publications department will be responsible for the reporting, and a technical editor will be assigned to supervise the undertaking.

Some engineering and scientific organizations, both large and small, exist primarily for testing, research, and experimentation. Their main product is not the manufacturing of products, but instead, the production of reports and papers describing research procedures and results. In this endeavor, the technical writer is as necessary as the researcher.

There are many kinds of reports, depending on who will use the information being reported. External reports go outside the company to clients, government bureaus, and libraries. They become the basis for further research. Internal reports are written solely for use within the company. They may be service reports, progress reports, and maintenance reports, to name only a few.

The reporting skill of the technical writer is important to most technical fields because reporting is basic to the success of the enterprise. The techniques of reporting are the foundation of instruction books, technical papers, and various forms of promotion and publicity.

A basic approach to technical writing is suggested by Matt Young in *The Technical Writer's Handbook* (see bibliography in Appendix A at the back of this book). In his preface he states: "My prescription for technical writing is this: It is not (or should not be) any different from other writing." He presents a few simple rules:

1. Write the way you talk; then polish.
2. Write one thought per sentence.
3. Be explicit.
4. Write for the uninformed reader.

He continues by stating:

> Many technical writers, unfortunately, seem to forget that their intention is to communicate, and they write as if for themselves. Their papers are insufficiently explanatory, and they are written with little or no regard for style or clarity. Even relatively good technical writing is frequently characterized by long, complicated sentences and difficult prose.

He even points out that important papers may be ignored because no one can understand them, so the work about which they are written remains to be discovered independently by someone else!

## INSTRUCTION MANUALS

Every profession has its professional jargon. By this we mean that certain words get invented; they comprise a kind of language that saves time for its users.

One of the jargon words in technical writing is *software*; it is, in many ways, the opposite of *hardware*. A company may manufacture precision instruments, flight instruments for airplanes, or word processors. These are examples of hardware. They are frequently the end product of an operation. They can be used in a concrete, hands-on manner.

*Software* is often associated with instruction books and manuals of use. Whereas *hardware* is the *end product, software* usually is the *means* by which hardware is produced, used, and repaired.

Common forms of software are instruction books and manuals, which may be written by technical writers within a company or farmed out to communication companies and agencies. It is estimated that the writing of manuals and instruction books makes up at least one-third of the software production of a company. That doesn't mean that you, as a technical writer, will necessarily be writing instruction books. What it does mean is that of the more than five thousand technical writers in this country, as listed by the Society for Technical Communication, about one-third are engaged in preparing manuals, doing the actual writing and editing and operating in supervisory capacities.

In discussing the job duties of technical writers who are involved in the production of instructions, it will be helpful to call upon the expertise of S. J. Goodman. Mr. Goodman has gathered his experience as a manuals writer into a pamphlet entitled *A Guide to Instruction Book Preparation,* and he has given us permission to quote his remarks as we examine how an instruction book is put together:

> An instruction book is informative literature necessary for proper support and presentation of a product. It may be considered as a permanent field representative of the com-

pany—often aiding and reinforcing the services of field engineering personnel. An instruction book is used by the operator of the product and by the service technician. It describes the product and contains information on installation, operation, theory, and maintenance. The publishing of a quality instruction book to coincide with engineering development and factory schedules requires coordination and cooperation between the sections of the technical publications department and their members.

Mr. Goodman goes on to say that writing is only part of the procedure in putting together any kind of publication. Even before an instruction book can be started, a number of steps must be considered, and the duties of each staff member must be well defined:

1. *Manager.* When the writing of an instruction book has been authorized, the manager reviews what is required, issues a project work order, and assigns various people to work on it.

2. *Writer.* The writer prepares a preliminary outline of the instruction book based on the specifications handed to him. He will include not only what must be written, but also a proposed list of illustrations.

3. *Manager.* The manager calls a conference including the writer, illustrator, copy editor, production editor, and if possible, the project engineer.

4. *In Conference.* All of these people consider the following:

    The scope and contents of the outline.

    The date on which the equipment the instruction book is supporting must be delivered.

    Where the instruction book fits into the overall schedule for all publications.

    Existing workload in the department.

    Amount of work required to complete the instruction book.

    Time required to have it printed and reproduced.

    The deadline date for each section of the book.

5. *Manager.* The manager informs the person or department who originally ordered the instruction book when it can be delivered.
6. *Production Editor.* The production editor draws up a schedule for the work to show how the various people will contribute to it.

Who are the staff members we have been mentioning? The production editor is in charge overall and sees that the particular instruction book is worked on, completed, and delivered on schedule. Copy editor is another name for the technical editor who edits the written part of the instruction book and, in general, reviews it for style, accuracy of expression, grammar, and punctuation. The product engineer is in direct charge of the device or system for which the instruction book is intended.

You are probably most concerned with what the writer does when handed an assignment to put an instruction book together. The procedures may be listed as follows:

1. Collect and study available production drawings (schematics, wiring, and assembly diagrams). Obtain related written information already existing (development reports, test procedures, and instruction books on similar equipment). If possible, obtain the equipment for further study.
2. Meet with the project engineer and other responsible personnel in order to obtain additional data.
3. Based on information from 1 and 2, review the preliminary outline and list of illustrations. Revise this preliminary outline to make a working outline.
4. Deliver the list of required illustrations to the illustration section and discuss how they will be produced.
5. Begin writing the instruction book. During the writing procedure, prepare sketches, revise existing production drawings, and describe requirements for illustrations. Identify illustrations by figure number and title and forward them to the illustration section.

6. As writing continues, the illustrator will submit roughs of illustrations for preliminary check.
7. Make arrangements with the photography section for necessary photographs.
8. Review completed draft. Edit copy for technical accuracy, format, content, and correct references to illustrations and paragraphs.
9. Submit corrected draft to the editorial section for preliminary review.
10. Following conferences with the copy editor, prepare final draft and forward it to the editorial section.
11. After receiving the reviewers' copies of the final draft, check and incorporate reviewers' comments into the master copy.
12. Review the manuscript completely, rechecking all illustration and paragraph references, format, and paragraphing for technical accuracy.
13. Send the completed manuscript to the editorial section.
14. When galley proofs have been received from the printer, review carefully, correct, and forward the corrected galleys to the editorial section. The importance of careful proofreading cannot be overemphasized.

In any publications department that prepares instruction books, you can see that the writer is really involved in the following four phases:

- *Research.* The writer analyzes the requirements of the project, collects the preliminary data, examines and interprets the data, and prepares the outline.
- *Development.* In the development stage, the outline prepared in the research stage is used by the writer as a guide in writing the text and in determining the illustration requirements of the first draft.
- *Prototype.* The research and development stages produce a prototype or model of the instruction book.

- *Production.* In the production stage, all effort is concentrated on mechanically producing the book.

These, then, are the specific duties of the manuals writer; but in terms of procedures, they are fairly representative of any kind of job the technical writer may tackle.

## PROPOSALS

As we mentioned earlier, proposals are among the most important forms of technical writing. They constitute a special kind of sales document. It is the rare technical writer who doesn't get involved in a proposal sooner or later.

Government agencies, frequently branches of the military, as well as private industries and foundations, regularly solicit proposals from suitable companies to conduct research on a particular problem or to design a mechanism or facility. These "Requests for Proposals" may involve anything from investigating the socioeconomic impact of a new manufacturing facility—or assessing its environmental impact—through developing new traffic control patterns. They may even involve the design and manufacture of new military equipment.

Companies responding to "Requests for Proposals" must design proposals, often several volumes long, which convince the government agency or industry that their company's suggested research project or design work best fits their requirements and is worthy of funding. Often these proposals must be submitted within a short time after receipt of the proposals request, so expertise in proposal writing is essential.

Proposals may seek funding in the millions of dollars. A company's success and its continued existence depend upon having its proposals accepted and funded. Without funded projects, there is no new business and no reason for the organization to continue to exist. Academic research units are in much the same position; without funding obtained through proposals, they cannot continue their work.

In all companies involved with proposal preparation, the technical writer serves a crucial function, and the ability to write concrete, persuasive proposals within tight deadlines is a highly marketable skill. Later, of course, these same writers may be involved in the preparation of the report and impact statements growing out of the research projects.

Regardless of the kind of proposal involved, it is a vital ingredient in all companies, large or small, in government agencies, and in universities. So, if you learn that General Electric or Lawrence Livermore Laboratory or Lockheed has received grant money, you may be sure that a proposal writer was involved in obtaining the grant.

## TECHNICAL WRITING FOR COMPANY MAGAZINES

Your first job as a technical writer may be with a company magazine, such as Raytheon's *Electronic Progress.* Sometimes the table of contents of this publication contains a wide variety of articles on electronics. At other times, an issue is devoted to discussing a particular field in depth. In any case, you would probably be writing some articles of your own or editing other people's articles.

You may ask, why do engineers and scientists write articles? The prime reason is the communication of knowledge. They may be engaged in research, may have developed a new technique, or may have been called on to publicize something for their companies. There is also the prestige factor—it enhances the author's professional reputation to have an article accepted and published.

In working on a company magazine, you will be helping others prepare their articles. You will find that most magazines consist of three departments: editorial, advertising, and production. As a writer, you will be assigned to the editorial department, but if you should demonstrate advertising or production ability, you may be able to transfer to these departments.

One of the best features of working for company magazines is that they are likely to operate with small staffs on a fairly informal basis. This often results in interesting working conditions, for you could find yourself doing a variety of things: writing original articles, editing articles written by other people, editorializing, and carrying out special assignments.

Sometimes a company magazine can be published by only one or two people working with a printing firm and outside advertising people, or it may be produced in-house with the use of computers and laser printers. It can have a much larger staff, including editors, proofreaders, copy readers, illustrators, office support staff, and an editorial board.

Working on a company magazine will bring you in contact with many people; this will require patience and tact on your part. In the first place, you must persuade the engineers and scientists to write for you, and they can be very busy people who often are not particularly interested in whether they get published.

Most editorial boards start preparing an issue of a magazine by having a "think session." And from here on, many of the things said about company magazines can be equally applied to commercial technical magazines.

A think session occurs when the magazine's staff gets together to decide just what is to go into a particular issue. And you may be surprised to learn that magazines work six, seven, or eight months in advance. It takes weeks and weeks to produce an issue from the first article until it is wrapped up and sent to the mail room.

One of the most fruitful sources for papers and articles is conferences attended by company employees. Papers delivered at such gatherings often turn up as magazine articles. But such selection is made with several factors in mind. The editor of the magazine may find it necessary to consult the public relations department to determine what is going on in the company that will appeal to readers. In addition, management must be consulted, because in the long run, responsibility for the magazine rests with the company's administrators.

Once you have received a manuscript from an author, the real job of writing and editing begins. A good deal of the experience you have gained in school, particularly in your writing classes, will come into play in this part of your job. For while the people from whom you get the material may (or may not) know how to write, frequently they may not have paid attention to their potential readers. Determining the audience, then, becomes your job. You must fit the style and tone of the piece to the audience you are trying to interest. For example, you may have to write a completely new opening for an engineer's paper. Here is where your job is perhaps most satisfying—and sometimes most frustrating—as you try to weld unorganized material into acceptable form. Changes of any sort, of course, necessitate conferences with the author, who must be as satisfied with the final product as you are.

The manuscript must be reviewed to be sure it fits the prescribed editorial format. It is then turned over to the appropriate staff people for illustration and layout.

## TECHNICAL MAGAZINES

We have given you a fairly detailed account of what it is like to work for a company magazine. But there is also a wide range of technical and professional magazines. Technical magazines are commercial and are published in order to make profits for their owners. Some examples are *Product Engineering, Machine Design,* and *Chemical Engineering.* These magazines are always on the lookout for accomplished young men and women to join their staffs.

Professional magazines are published by professional societies. Some examples of these are *The Journal of Chemical Education,* published by the American Chemical Society; *Civil Engineering,* by the American Society of Civil Engineers; and the

*American Journal of Nursing,* by the American Nurses' Association. In an issue of *The Chronicle of Higher Education,* the Mathematics Association of America advertised for an Editorial Manager to work in Washington, D.C. The duties of the position included supervising the editorial department for three journals and several books a year. Training in mathematics, excellent writing skills, and some editing experience were required.

A great many journals are read by audiences outside one's company. For this reason, if you become a technical journalist, above all you must be able to gauge the interest and needs of these external readers and to produce material which is "reader friendly." The ability to analyze these readers becomes of paramount importance. If you feel that you have this ability, together with imagination and proper motivation, technical journalism may be your career goal.

## TECHNICAL PROMOTION

A brochure put out by a major company starts this way:

> Babcock & Wilcox is a major industrial company employing thousands of people in the manufacturing and marketing of highly engineered, essential industrial products.

Another paragraph continues:

> The company's beginning dated back to 1867 when Babcock & Wilcox started with two men and an idea. In that year, George Babcock and Stephen Wilcox developed a new and safe boiler to provide power for a rapidly industrializing world. B & W evolved steadily until now, well over a century later, the company ranks among the largest industrial corporations with annual shipments in the range of a billion dollars.

These words are from a recruiting bulletin designed to attract well-trained people to work for the company. It is fairly typical of what is called technical promotion, that is, writing specifically geared to selling a company and its products.

Let's examine the following hypothetical situation to illustrate this point.

Suppose that the Allied Electronics Company is going to conduct research on transistors for the government. The research and the time consumed in the laboratories will be only a part of the project. One preliminary task will be to obtain the necessary laboratory equipment from other companies, such as the Eastern Instrument Company. And that may mean everything from switches to the most complex instrument panels.

Allied Electronics, then, must be aware of the current product line of Eastern Instruments, their reputation, and the quality of their products. In addition, they must be assured of their ability to modify the design of items they manufacture.

This is where the technical sales writer for Eastern Instrument comes in. Advertising copy for technical magazines must be written, and sales literature must be prepared for prospective customers, such as Allied Electronics.

It is almost impossible to classify the various types of sales literature, but let's suppose that you, as a technical writer, are engaged in preparing a sales brochure.

The procedure is about the same as for other pieces of technical writing. First, the project itself must be authorized, in this case, by management. Then, a number of things take place, sometimes concurrently.

You will first collect all the information about the product or equipment that you possibly can. You must become totally familiar with the background of the project. For this, you may have some earlier booklets to serve as guides. You will visit the departments responsible for the design, construction, and manufacturing of the product to get its views on the consumer, the kind of company it is, and anything else that will help the booklet put across its message.

A technical writer working in sales literature must be aware of how much money can be spent on the brochure and how many copies are going to be distributed. These two factors will sometimes determine whether the brochure is to be printed within the publications department or given to an outside printer.

The technical writer then designs the brochure, working in close cooperation with a designer. This is where this kind of publication differs from some others. In sales literature, the layout of the brochure is extremely important, involving questions of shape, size, color, and illustrations.

The copy in an advertising brochure is frequently subordinate to the illustrations, and the writer must decide how much copy to write to support the illustrations. At this point the actual writing of copy begins, followed by checking and revision, and all the other stages that go into any industrial writing.

## NEWS RELEASES

Another kind of technical promotion is the news release. This may be in the form of a news story or an article for a magazine. The real reason for preparing a news release is to supply information (and advertising) that editors will want to include in their publications. It must be carefully written to present the information clearly and concisely, with language chosen to interest and impress the editors to whom it is submitted.

The technical writer often gets involved in news releases, operating from either a regular publications office or an advertising department. That there is considerable skill involved in preparing news releases is pointed out by Sydney F. Shapiro, managing editor of *Computer Design Magazine.* Mr. Shapiro says:

> The most newsworthy item may be ignored if the presentation to the editors is poor. The true measures of the success of a news release are how many times and where it gets into print.

> To assure consistent acceptance of his news releases, the writer must know the interests of the particular editors and

of the magazines' readers; must prepare valid, newsworthy releases; and must submit the releases in the proper, easy-to-use format that facilitates their use.

For individual readers, the news release often comes in another form called new product information. This variation must be short and to the point. It should demonstrate confidence in the product, describing it briefly and requiring a minimum amount of space.

This is an example of new product information:

> This Hand-Held Anemometer will easily measure air speed wherever your hand can reach. The accurate hand-held one-piece unit weighs only three ounces and requires no external probe. Running on sapphire bearings, its freely turning turbine will rotate at a speed directly proportional to wind speed. The rotation is passively sensed by an infrared light beam which adds no friction. An integrated circuit even converts the signal to your choice of units (feet per minute, mph, meters per second, or knots) and feeds it to a three-digit LED display. This cleverly designed instrument, which operates with three AAA batteries, measures 4-1/2 × 4-1/2 × 1-1/8 inches.

It has been estimated that daily newspapers receive from 25 to 250 news releases a day. Approximately one of every 25 is used. As you can see, dealing with news releases requires special skills, many of them learned on the job.

## TECHNICAL ADVERTISING

Manufacturers of machines, instruments, and other industrial products spend millions on another kind of promotion. This is technical advertising.

Sargent-Welch Vacuum Products Company placed this ad in an issue of *R/D* magazine:

**A Pump for Every Vacuum Range**
You've got the vacuum requirements—we've got the pumps. Pick your own range and there's a Sargent-Welch pump right there ready to go to work—from the famous Duo-Seal oil-seated rotary vane pump line to the ultra-high, ultra-clean vacuum Turbomotor line of turbo-molecular pumps with capabilities to 1,600 liters per second. Or maybe one of our New Director direct drive pumps will fit your application better.

What we have given you is only the copy, or description, written by a technical writer either in the company's advertising department or in an agency hired by the company. The ad also features illustrations, various kinds of type, and other attention-getting devices.

It should be apparent that to write copy for technical advertising you must know something about the technical part of the product. You should also know enough about graphics, illustrations, and charts to give directions to the printer. And you should have some familiarity with composition—how various items are positioned on the page.

So, as a technical writer in advertising, you may work in the advertising department of a company or you may get a job with an outside agency specializing in technical advertising. In either case, your job duties and projects will be similar. One private advertising agency lists the following specialties:

Market research
Program planning
Publication advertising
Direct mail advertising
Publicity
Technical literature
Merchandising aids
Slide presentations

The writer of technical advertising may be involved in institutional advertising, for example, by writing general copy that creates a favorable image of the company in the public mind. New products must be advertised or old products changed to such an extent that they seem new. The advertiser may also call attention to the service and maintenance offered by the company.

Robert D. Towne, president of W. L. Towne, Inc., an advertising agency in New York City, has outlined several points which help to explain the duties of the technical advertiser.

First, he says, advertising writing is different from other kinds of writing because it is persuasive. In other words, even though information is at the heart of advertising, its main purpose is to persuade people to buy a product, a service, or to have a problem solved. To many writers, this offers an interesting switch from the run-of-the-mill technical writing.

Additionally, the technical copywriter must think not only in terms of writing but also in terms of two other factors: the sales idea and illustration. These will bring the writer in close contact with the sales force of a company and provide the stimulating experience of working with fine illustrators.

Technical advertisers also have their think sessions, as ideas are tossed around for review and the objectives of the advertising campaign are discussed. Dozens of ideas will be looked at and discarded, but somewhere will be the one that will please everyone, especially the client.

## TECHNICAL FILMS AND VIDEOTAPES

The technical or scientific film can be an effective way of selling a company's service or product. Film and videotape play an increasingly important role in the training of technical and nontechnical employees. This is another area in which the trained technical communicator can find worthwhile employment. Scriptwriters do most of the technical writing in the production of films and tapes. This position can be interesting because it involves a multi-faceted medium—communication linked with visual aids. Technical films are used for a variety of reasons:

- A film or tape cuts across many audiences. It may be interesting to a large group of people and yet, at the same time, have a more specialized appeal to a particular smaller group.
- It can do all kinds of things that the product by itself may not be able to do and that a still photograph can do in only a limited way. Through drawings and cartoons, the film can enlarge views, reduce them, allow one to see inside a device, linger over it, and repeat it—all in motion.
- Films and videotapes have impact. Research has revealed that the combination of sight and sound impresses ideas and facts most emphatically upon the audience.

The scenario—the manuscript containing the words you will ultimately hear on the screen, together with action of all kinds—begins once a central idea for the film has been determined. The scriptwriter must decide what kind of audience is being targeted—whether it is a group of specialists, a group of managers, or people who are unfamiliar with the subject. This is a most critical stage for the technical scriptwriter.

Motion picture production is a complicated and costly procedure. If you decide to pursue a career in scriptwriting, it would be a good idea to get some training in this kind of writing in high school or college. And, although it isn't absolutely necessary, the beginning scriptwriter will certainly benefit by knowing something about photography. Courses in cinematography would be advisable as well.

The writer will have to visit shooting sites to become familiar with the location to be described in the script. And just as with any other writer, the product must be understood in depth. Things must be described in visual terms with a knowledge of what photography is capable of doing.

The story line is usually developed first. This is a kind of synopsis, or highly concentrated version of speech, action, and narration. Next, the actual motion picture shooting begins, scene by scene. Then the film will be reviewed, and the scriptwriter may be asked to write a narration, which is an accompaniment to what the action means, usually spoken by a professional actor or reader hired for the purpose. Words must be written which are easily spoken and understood and which synchronize with the photographed action.

Videotape also plays an important role in industry and offers opportunities for the technical writer with some understanding of the medium. In a series of articles, published in issues of *Technical Communication,* William Thomas predicted greatly expanded use of videotape recorders and computers, and suggested that the electronic visual media will be used to produce training programs. The technical writer of the 1990s must be aware of the possibilities offered by the computer in graphics production, as well as word processing.

## TECHNICAL TRANSLATION

> When Americans hear the phrase "technical writing," they normally think of writing in American English for an American audience about American technology. But technical writing goes on all over the world. . . . A growing number of technical writers must cope with translations— read them, evaluate them, sometimes (sadly) retranslate them from crippled English into sound prose.

These are the words of Ben and Frances Teague in an article in the *Journal of Technical Writing and Communication.* They point up the importance of technical translation, a rapidly growing development in technical communication.

If you can write in Spanish, or in German, French, or a major Asiatic language, you may find a job waiting for you in technical communications.

The importance of translation as a career can be seen every day if you read the science or business sections of the *New York Times,* or the *Wall Street Journal.* If you are living anywhere near an engineering library, look over the technical magazines. You will find articles that have been translated from foreign languages into American English.

Businesses and industries these days are global in nature. Most of the larger American companies could not exist in their present form if they weren't able to communicate with companies and people in other countries. The electronics, chemical, and transportation industries, and computer hardware and software manufacturers, among others, need manuals, reports, research papers, and technical advertisements which can be translated into foreign languages.

Grace Tillinghast, a technical translator in the International Photographic Division of Eastman Kodak, has said:

> In order to reach different areas of a world market, a multinational company has to publish in various languages. A product will sell better if the potential customer knows he or she will be able to read [in his or her language] the enclosed material. I came into the Kodak picture to help satisfy the needs of our Spanish-speaking customers.

As you look forward to receiving training as a technical writer, keep this advice in mind. If you have any facility in a foreign language, continue to develop it. You never know when it may be a skill that appeals to a prospective employer and will give you an advantage over the other applicants.

## DOCUMENT COORDINATION

Many technical writers and editors move beyond technical writing into document coordination. In this role, which is frequently a management position, the former technical writer is responsible for the entire document production process, following each document from the initial meeting with the client—in which the document's specifications are determined and the various activities of the researchers clarified—to the final publication and presentation to the client. The document coordinator is also involved in any modifications of the document in response to suggestions from the clients. This function demands all of the interpersonal and managerial skills of the former technical writer/editor. Document coordination requires the ability to elicit material from the technical staff (who are often reluctant to write their research results), to interact with clients who may be uncertain of their actual needs in a particular document, and to work within personnel and budget constraints imposed by the company's administration.

The document coordinator is often required to visit job sites, help with data gathering and analysis, monitor the production of graphics, perform public relations functions on behalf of his or her company, control production costs, and perform a multitude of other activities. You may think these duties are beyond the scope of technical writing, but they quickly become great sources of challenge and satisfaction to the able individual.

## TEAMWORK IN TECHNICAL WRITING

During their careers, technical writers come in contact with the management, research, and production divisions of their companies. Such contact emphasizes the fact that the technical writer is part of a team.

You may now be asking, what is meant by teamwork? In brief, it means contributing your skills as a writer or editor to a project at the same time that other people you are working with contribute theirs. Here are a few examples:

1. You will undoubtedly have to work with nonwriters. By this term, we don't mean uneducated people; we mean, simply, people who may do very little technical writing throughout their careers. They may be engineers, managers, personnel people, artists, or audiovisual experts. Always remember that they are experts in their fields. Sometimes you may have to call on these people to provide the basis of a report or a research project, or to suggest an illustration to go along with an important piece of work. Sometimes these inexperienced writers will give you sketchily produced work: the sentence structure may be poor or confusing; the style may not be your idea of good writing; the grammar may violate what you have learned in school.

   Here is where you act as part of a team. You play down any feeling of superiority you may have acquired. You will soon learn that the illustrator is a far better illustrator than you will ever be; the researcher is a far better chemist than you will ever be. Each contributes his or her special knowledge and experience to produce a successful project.

2. You will frequently work with people who have given little thought to the kinds of readers to whom they are supposed to appeal. In this case, it would be up to you to ask the author of a scientific paper you are editing such questions as: how much do your readers already know about the subject; how "technical" must your terminology be; do you have to spell out everything? These are the questions of a teamworker, and they must be asked with much tact and a great deal of consideration for people's feelings.

3. Lawrence T. Hammond, a research writer with Halliburton Services, brings up another situation where teamwork is essential. In brief, he says there is a phenomenon in communications management that many consider a mystery. What causes barriers in internal communication—that is, communication within a company? Some companies seem

to spend a lot of time trying to get people to talk to one another pleasantly and profitably, yet they do not always succeed. This is a case for you to use whatever skills of cooperation you possess, to merge with the team, and to practice what we all mean by "communication."

Suppose you have been given the job of coordinating a proposal that is submitted to NASA. Your company thinks that it can produce a superior electrical system for a missile, and NASA is definitely interested in hearing about it. It is up to you to get all the technical data on the electrical system from the engineers who have designed and tested it. Their data probably will be in the form of reports, test sheets, and innumerable calculations. Your job here will be to take this mass of information, sort it out, and retain enough to impress upon the NASA people that your company is well qualified to handle the job.

But this isn't the end of your assignment. You will have to attend meetings of the sales force so they can help you put your proposal across, giving it sales appeal. Then somewhere in the process, probably after you have written the first draft, you will have to sit down with your company's top administrators, who will want to scrutinize your proposal backwards and forwards, inside and out. They must pass final judgment on what you have written, for the reputation and strength of the company depend on your effort, at least in the eyes of NASA.

Or suppose that you have been assigned to write a manual. This can be a big job, requiring a team of four or five people—a technical writer, an engineer, a designer, and an illustrator. Writing a manual is usually a long-term project. Although the material will originate with the engineers who worked on the equipment the manual describes, the technical writer must consult many other people as well.

You may function as the person in the publication department who is responsible for the production of articles and papers. The basic material will come from the engineers and the research people, but you will work closely with the public relations department in trying to place the articles in national magazines. To get a single article in shape, you may meet with a veritable barrage of management executives, patent lawyers, and supervisors of one kind or another.

You can see, then, that the technical writer is not isolated in a tiny cubicle somewhere. Hours must be spent in writing, and many hours in preparation. The writer becomes adept at interviewing and attending meetings and becomes thoroughly familiar with the divisions within the company.

A lot of the variety in technical writing is illustrated by two definitions supplied by B. H. Weil of the Technical Information Division of Exxon Research and Engineering Company:

- *Technical Writer:* Writes technical reports and articles for specific audiences, usually based on existing reports and on information obtained directly from personnel involved. Typical documents may be highlights of research progress reported by many groups, slanted for management information; overall status reports; and technical papers or chapters of books, where expedient.

- *Technical Editor:* Responsible for either expediting or managing the writing and production of reports and papers of the types desired by management, in the forms and styles agreed upon. Edits rough drafts prepared by the technical personnel themselves—before, during, or after supervisory review, as desired. May coordinate illustration, proofreading, printing, and initial distribution. May prepare or coordinate style manuals and technical writing courses.

Mr. Weil speaks of illustration, printing, and distribution. He points out that a typical publications department consists of editors and writers, photographers, illustrators and other graphics people, reproducers, printers, and production people. The interaction of many people is required to prepare a publication for print. For maximum effectiveness in transmitting technical intelligence, consider the concept of writing-illustrating teams: not art as an afterthought, but complete integration of effort from the inception of a task.

Here is one example of team cooperation in the publications department. In the production of a report, the technical writer must first obtain a rough draft from the engineers. Then there are conferences with them, pointing out where the structure is faulty, where the style can be improved, and where changes have been made. All of these changes must have the approval of the original writers—the engineers themselves.

Finally comes the day when the manuscript is ready for reproduction. But in the meantime, the technical writer has conferred with each supervisor, and has consulted the graphics department about pictures of the research involved or arranged to have some sections of the apparatus drawn. Contact has been made with an illustrator or graphics expert to have a chart or graph made.

Someone in the printing shop must be consulted, if the material is to be printed. Whether the material is to be printed or reproduced by some other method, schedules with the appropriate technicians must be made. Deadlines must be set with all of these people, a time when all the pieces will be put together and will be ready for final printing. If the schedule has been planned to allow enough time, there will be less chance of a bottleneck in getting the final report in the mail. The pressure of final deadlines can be formidable at times, and the ability to function under them is a prime requirement for technical writers and editors. The proposal or report you are preparing may mean many thousands of dollars worth of business to your employer; the deadline on such a project is not likely to be taken lightly by the company's management.

# CHAPTER 3

# EMPLOYERS OF
# TECHNICAL WRITERS

In the previous chapter, we discussed the different types of jobs technical writers might perform. Now let's look at some specific employers and their technical writing staffs.

But before we get into specifics, we refer you to a few statements from a survey of publications managers throughout the United States about the depth and breadth of publications in this country.

These are some of the conclusions offered:

Most technical publication work is done in industrial multidivision corporations.

Most publication work comes from organizations with over 2,000 people, and the publications departments report primarily to engineering, administration, and marketing.

Most job titles are technical writer, technical editor, and technical illustrator.

An organization that requires personnel with degrees will hire those with a B.A. or B.S. in technical communication and others with a B.A. or B.S. in English or journalism. The vast majority of managers believe that employees with degrees in technical communication are better prepared.

There are many more items in the survey, conducted by the Society for Technical Communication, than those we have quoted here.

The employee requirements of technical communicators are also highlighted by Clinton Hawes, a staff communication specialist for IBM. He says that the diversity of skills needed is indicated by the many types of technical writing done in many companies.

## INDUSTRIAL COMPANIES

Almost all government research contracts contain a clause which requires that industrial companies provide periodic reports of progress. This points up the need for technical editors as liaison between research and administration. To quote an authority on technical writing:

> Today, government contracts account for as much as 90 percent of the total business of many of the larger companies. Because of these contracts, the volume of progress reports, correspondence, and interplant communications has expanded enormously. The result has been that since 1945, industry has attempted more and more to employ professional writers with scientific backgrounds in order to take the load of product explanation off the already overburdened engineers.

IBM exemplifies industry's heavy reliance on technical writing staffs. IBM is a very large company, manufacturing a wide variety of products in plants all over the world. IBM employs technical writers in each of its plant locations—to write reports which are sent from one department to another. For example, they are circulated between laboratory and top management, and from domestic to international branches to prepare sales literature when the company puts such products as microcomputers and office equipment on the market; and to propose new ways of handling great masses of technical information.

IBM has offered many opportunities to young men and women to succeed in the world of technical communications. One of these has been Anthony J. Sammartino, who, after he was demobilized from the U.S. Army in Europe, started and organized the publications department at IBM's Stockholm facility. With his fluency in several languages, Mr. Sammartino became a manager in the Far East Corporation of IBM.

The electronics industry employs many technical writers. One of the first to feel the need was the Northrop Corporation. If you were to work for Northrop, one of the divisions to which you could be assigned may deal with avionics. This is a comparatively new word, but in ordinary terms, avionics refers to those parts of an aircraft concerned with how the plane gets from one place to another, how the flight crew can communicate with various planes, and how the plane is controlled while it is in flight. Because it is so technologically complex, a great mass of reports and papers is generated by technical writers in this branch of electronics.

The publications staff at Northrop became very large shortly after World War II. Today, it numbers well over 100—with supporting personnel consisting of illustrators, parts catalogers, word processor personnel, and others.

IBM and Northrop are only two of the many industrial companies throughout the United States which employ large numbers of technical writers.

One of the areas in which these companies use technical writers is in internal communication. In such large companies, it is essential that the employees on all levels and at all locations be kept informed about what is happening in their company. This information is circulated by a variety of communication techniques such as in-house newsletters, FAX machine memoranda, and internal group meetings.

General Electric Company provides its scientists and engineers with a wide range of supporting services. The need for technical writers and editors at GE is demonstrated in this statement:

> General Electric is one of the largest single sources of technical papers for engineering and scientific journals. Employees are encouraged to publish technical work, but at the same time are relatively free from pressure to prepare routine reports. . . . General Electric scientists and engineers are encouraged to attend and participate in meetings of professional societies.

Implicit in this statement is the role of the technical writer, relieving the company engineers of routine reports and aiding in presenting articles and professional papers.

Technical writers have also found a secure niche in the chemical industry. A technical writer in Allied Chemical Corporation describes a publications job as editing technical and safety analysis reports, describing design criteria, and helping to write and edit journal articles, papers, and brochures.

To illustrate the duties of technical writers working for such companies, it might be helpful to refer to a job description from one of the chemical companies. This company's technical writers are expected to:

- Look up in journals, magazines, and the publications of other companies the technical literature that would be useful to the service and sales groups.
- Direct the preparation of rough drafts by those technical people responsible for developing new products for the industrial market.
- Edit copy and supervise the layout and printing of technical literature.
- Prepare articles for technical magazines and speeches for technical conferences.
- Coordinate the total literature output of the research departments.

The idea that only large companies employ technical writers is incorrect. With the increase of small companies as subcontractors, the volume of paperwork has increased greatly. Small companies are not necessarily selling their products to technical companies; many of them sell directly to the public. Because consumer products are increasingly complicated to operate, even small companies are compelled to furnish well-written instructions with their products. So our recommendation to you is to consider a variety of companies of all sizes when you are seeking employment. Remember that the directions for do-it-yourself kits for household equipment had to be written by someone.

At the same time that you are thinking in terms of smaller companies, consider other places for employment. Some technical writers prefer to work for agencies producing services rather than goods. Technical writers have worked for such agencies as the Travelers Insurance Company in the capacity of senior technical writer in the engineering division. Another agency known to hire technical writers is the National Oceanic and Atmospheric Administration.

## RESEARCH GROUPS

Heavy industries, such as those mentioned above, find technical writers indispensable. Another source for technical writing positions is with research organizations. Some are part of the companies themselves, some are supported by universities, and others are privately endowed.

Regardless of their management, the research groups have one thing in common. They are looking for new ways of doing things, new products, new communication systems. Frequently their efforts do not show results for many years, but they generate an incredible flow of reports, science and engineering papers, and presentations before scientific and engineering audiences.

Think of the exploration going on to obtain new sources of energy, whether in oil, coal, or solar power. One organization involved in energy research is the Jersey Production Research Company. Jersey Production Research, an affiliate of Standard Oil Company, is concerned with better methods of manufacturing old products and ways of producing new ones. To quote one of its brochures:

> Most fields of physical science and engineering are represented in our employees' backgrounds. You will enjoy working with these people and will benefit from the experience. Our people come from ninety colleges and universities, including schools in Paris, Berlin, and Bern. On the basis of their highest attained degree, 42 percent of our professional people have Bachelor's degrees, 27 percent have Master's degrees, and 31 percent have their Ph.D.

If you were a technical writer for Jersey Production Research, you would be associated with engineers and chemists trying to improve current methods of extracting oil from underground rock strata; with electrical engineers designing special instruments for oil exploration; with physicists studying how sound waves travel in the earth, the patterns of electric currents, and the changes in the earth's gravitational field; with mechanical engineers trying to determine the oil-producing capabilities of geological formations; and with mathematicians using high-speed computers to predict how well and how much oil pools would produce. As a technical writer, you would be in a publications department geared to support all this new and exciting research.

Some research organizations are not affiliated with large industrial concerns. A special type of the research institute is exemplified by Battelle Memorial Institute. Battelle contracts with both industry and government to develop scientific studies of many kinds on its own. It is not a consulting firm or a testing laboratory; it is devoted to research, with a staff divided into approximately fifty operating divisions. To name these divisions would take too long, but some that are representative perform research in astronautics, communications science, environmental science, environmental systems, gas technology, cancer studies, lubricant technology, paper technology, textiles, welding, and forest products. You can see that the scope of activity for a technical writer associated with Battelle is almost endless.

Many universities also have large research organizations that are heavily dependent on government contracts. One of these is the University of Dayton Research Institute; another is Lawrence Livermore, affiliated with the University of California.

## GOVERNMENT AGENCIES

Government units employing technical writers usually fall into two categories: federal and state groups that use technical writers for their own work, and agencies that work closely with companies and industries.

You should take a look at the directories put out by the United States Civil Service Commission, in the section entitled "Federal Careers." This publication contains two descriptions of federal careers for writers:

- *Public Information Specialist.* In this category are writers who not only collect information about the many activities of the Commission, but who also write and disseminate information about the many programs available in federal government. As an information specialist, you could be involved in writing for a variety of public communications media, including newspapers, television, and magazines.
- *Writer-Editor.* This job should appeal to you if you have substantial knowledge in the areas of engineering or science. The federal civil service employs writers and editors to produce articles, press releases, periodicals, pamphlets and brochures, speeches, and scripts for radio, television, and film.

As a writer-editor for the federal government, you would research the subject to be described, select the information to be included, and write or edit the final manuscript. Many of the writer-editors in this group specialize in technical fields such as engineering, science, or the social sciences.

We have just described what may be called internal information in government agencies. But there is a second classification of government writer whose job involves strong industrial contacts. Some government agencies work so closely with private companies that it is hard to distinguish between the two. One example of this relationship is the Mound Laboratory at Miamisburg, Ohio, operated for the U.S. Department of Energy by the Monsanto Research Corporation.

Mound Laboratory is in the forefront of energy research, providing leadership in such areas as polonium technology, thermal diffusion, and reactor fuel studies. Because its research is so complex and extensive, Mound Laboratory has a need for many skilled technical writers. One writing group is responsible for the preparation of the manuals which must accompany every project before its results can be implemented. Another group is the Technical Information Office, which is responsible for the preparation of technical papers for publication in journals, for answering inquiries of a technical nature which are received by Mound Laboratory, and for editing and publishing periodic and progress reports.

Government research groups are not necessarily run by large companies. Some government agencies are to be found in the military itself, developing weapons, missiles, and equipment for space exploration. Harry Diamond Laboratories, part of the U.S. Army, is one of these agencies, as is the Naval Weapons Center at China Lake, California. These military agencies provide numerous career opportunities for civilian and enlisted technical communication specialists.

## JOURNALS AND MAGAZINES

As we have discussed previously, the technical writer who is working in the journal field is usually required to edit someone else's work and prepare articles for publication. In doing this, the writer works closely with the author, restructuring ideas and checking grammar, punctuation, and spelling. Eventually, the technical editor may be asked to write original articles in some specialized fields. A brief discussion of the different categories in the technical magazine market should help illustrate the variety of career opportunities available.

First are the journals, sponsored by professional societies. You are probably familiar with a number of these—you may even belong to a chapter of an engineering society, such as the American Society of Mechanical Engineers.

Practically every professional association publishes its own journal. *Aerospace America* is self-descriptive, as is the *American Journal of Agricultural Economics.* Two of the better known ones are the *Journal of Chemical Education,* published by the American Chemical Society, and *Engineering Times,* the journal of the National Society of Professional Engineers.

These journals have several common features. They usually publish papers based on original research; they operate with comparatively small staffs; and they are mainly read by people in the same field as the sponsoring society. Regardless of its individual makeup or audience, the technical journal must be edited by skilled technical writers.

One of our former students, for example, is presently the assistant editor of *Theriogenology,* an international journal of animal reproduction. She is responsible for editing (including visual aids) all articles submitted. Many manuscripts, especially some submitted by foreign authors, require extensive revision. She also indexes the volumes of the journal, compiles the front matter, and corresponds with authors and reviewers.

Commercial magazines are found in technical libraries in every country. McGraw-Hill publishes over thirty technical, scientific, and business magazines, including *Product Engineering, Electronics,* and *Power.* Another well-known technical magazine organization is the Penton Publishing Company, which produces *Machine Design,* among others.

Most editors of commercial technical magazines are interested in interviewing qualified technical writers for staff positions. These editors realize that in a journalism field as competitive as theirs they must inject new blood into their organizations and hire new people with good technical training and the ability to write. And they are willing to give them on-the-job training.

But you should realize that there are differences between working on journals and working on commercial magazines. The latter are money-making concerns, employing large editorial staffs. For this reason, the chances of obtaining a job with a McGraw-Hill magazine or with Capital Cities ABC, publishers of *Iron Age,* are greater than with such specialized publications as *Journal of Nuclear Materials* or *Separation Science and Technology.*

One particular form of technical magazine is the company magazine, often called the house organ. This is put out by the company's publications department, not by a publishing company. House organs usually fall into two classes: some are for outside readers, others for internal readers. The *RCA Engineer,* published by the Research and Engineering Division of RCA, is a highly technical publication. On the other hand The Oak Ridge National Laboratory publishes the *Review* largely for internal readership, and is distributed to employees and others associated with the Laboratory. The staff writes and edits a variety of articles: some deal with interesting people employed by Oak Ridge National Laboratory; others with work in progress in the research area. A magazine like the *Review* could offer an opportunity for a writer with a combination of training in technology and journalism.

Another class of magazine is the trade journal. It is a little difficult to define a trade journal; however, it bears the same relation to a technical magazine that a trade bears to a profession. It features down-to-earth articles on how things are done, methods of production, and tips to readers in various trades. There are trade magazines for persons who service television sets, cover floors, repair roofs, install large-scale boilers, and for those in other occupations.

Although we have been talking about magazines, we should not overlook the book publishing companies as potential employers. John Wiley and Sons is only one of several large publishers of engineering and science textbooks. These require editing by highly qualified people who act as liaison between the company and its authors. A graduate of Rensselaer Polytechnic Institute's Master's program in technical writing described his position with the McGraw-Hill Book Company this way:

> I am a staff editor with the McGraw-Hill *Encyclopedia of Science and Technology.* . . . It is a fifteen-volume work containing articles by top-flight people in all fields—we have some 3,000 contributors in all. . . . I am handling most of the field of physics, plus aeronautical and nuclear engineering and space technology, as well as other miscellany.

More and more publishing houses that produce technical and scientific books are looking for specialists—technical editors who can help authors and who are familiar with the content, vocabulary, and audiences of technology.


## SUPPORT COMPANIES

The last forty years have seen the development of small companies that might be called support companies. These support companies produce technical brochures, manuals, and other publications under contract with larger manufacturing firms. They are likely to fall into two groups: those that serve as consultants and help promote company products, and those that act as contractors by doing the actual writing.

One of the most active companies in the consulting field is Hall Industrial Publicity, Inc., of Pleasant Ridge, Michigan. When you read the biography of Stuart P. Hall, president of the company, you gain some understanding of what the company does and how you could prepare for similar work.

Mr. Hall, who holds a degree in mechanical engineering, eventually discovered that he had a talent for writing. He joined the editorial staff of *Product Engineering* magazine and later became associate editor. Then, after several other editorial jobs, Mr. Hall formed his own company to handle technical publicity, public relations, and technical catalog production for leading corporations selling their products to industry.

Mr. Hall's company deals in three general commodities: press releases, publicity, and technical articles for such clients as American Radiator and Standard Sanitary Corporation, Minnesota Mining and Manufacturing Company, and Bendix Aviation Corporation. He has on his staff a dozen or so people who are technical writers of a highly specialized nature.

Brief mention of a few other consulting and publicity companies will give you an idea of their variety and the opportunities they offer to the person interested in technical writing as a career:

Clark, Channell, Inc., a management consulting firm in Stamford, Connecticut, specializes in executive recruitment—helping other companies obtain personnel. As such, this company is primarily involved in recruiting technical writers. A job description from Clark, Channell for an editor-manager began:

> A nationally known research-consulting firm, specializing in human engineering and systems research, is expanding its in-house capability for editing and printing its own reports, proposals, and other written materials. Already

equipped for this purpose with a small, capable staff and modern equipment, this organization seeks a person who will supervise and further develop this supporting service group.

The role of the technical writer in preparing technical publicity, advertising, and other promotional forms is summed up by John B. Bennet of ITT Laboratories:

> The publications engineer-technical writer is trained and experienced in the preparation of publications to provide technical information to engineers, technicians, and administrators. He knows how to explain technical facts. He has good understanding of the company's products and, even more important, he has a feeling for the company's spirit and ways of doing business. He can, therefore, present technical achievements made by the company in the best possible light to the general public and to specific customers.

The line between consulting companies in technical writing and contracting companies is fairly fuzzy at times, but some distinctions can be made.

A contractor is essentially a specialized organization which relieves a larger company of publication responsibility when an assignment is received that is going to overload its publications department. It may not necessarily be an order; it may be a new development in that company which must be described, perhaps in the form of a report.

Suppose that Radio-Electronics Company has received a large order from the government for a fire control system on a line of Navy ships. Radio-Electronics is prepared and able to manufacture the system, but operating and maintenance manuals must be prepared. Rather than overtax its publications department, the company contracts with Roberts Technical Writing Service to prepare the necessary manuals. This outside company now adds Radio-Electronics to its list of clients for this job only. Perhaps it will be the only job on which the two companies will ever work together.

Several such contracting firms are spread over the country. Their existence depends on a number of factors:

- They are prepared to give specialized services in the preparation of catalogs, brochures, or training manuals;
- They can stick to prearranged schedules because they are not subjected to the pressures found in manufacturing companies;
- They can bring in part-time help for peak loads, a practice that a large company is reluctant to undertake.

A contracting writing service will plan the entire publication effort for a particular project, doing all the necessary writing and editing. It will offer a complete illustration and graphics service and will do the printing or see that it is done. It will deliver the final product—the manual, report, or catalog—to its client or distribute it to interested parties. In the long run, the contractor supplies a complete communications package to its client, with little responsibility on the client's part beyond the necessary input and final approval.

## TECHNICAL WRITERS IN HIGHER EDUCATION

Many colleges and universities are engaged in industrial research and development, particularly those with strong science and engineering faculties.

One has only to think of Stanford University's Research Institute and the Johns Hopkins Applied Physics Laboratory. However, groups such as these may operate almost independently of the sponsoring college. But there are hundreds of institutions throughout the country whose teaching staffs are active in government- or industry-sponsored research. As the number of academic discoveries and inventions increases, so does the need to convey information to industry and government, as well as to the general public. This policy of sharing results is known as "technology transfer."

An independent study of several such college research groups shows a trend toward hiring technical writers to prepare reports. The study asked a number of questions and received answers from a representative group. These are the results, compiled from the survey:

*Question:* Do you employ technical writers or editors?

*Answer:* Over half employ one or more writers. Two institutions not now employing writers are planning to, and those already employing writers plan to hire more.

*Question:* Are the writers who are employed working in public relations or in research?

*Answer:* Over half employ writers in public relations; a small number employ writers in research. Others employ technical writers in the university press.

*Question:* From what sources do you hire writers?

*Answer:* In some instances writers are graduate students studying communications. Others have been found in industries that have publications departments.

*Question:* What kind of training and experience do you require?

*Answer:* The answers were varied: five or more years of professional writing experience, a B.S. or B.A. degree, an interest in science and technology, and a flair for technical writing.

The results of this study show that there are many places in academic life for technical and scientific communicators. Joseph Sanders, for example, is a writer and editor cooperating with technical communicators at the Child Development Institute at the University of North Carolina at Chapel Hill. In addition to working with other technical communicators, he develops programs for assistance in using media devices.

The demand for people like Joseph Sanders may never be great. But the attraction of the university environment may make certain jobs in this area highly desirable to some technical writers.

## TEACHING TECHNICAL WRITING

More and more colleges are giving communications courses for students whose major fields of study are engineering, science, and liberal arts. At the University of Florida, for example, all future engineers are required to take basic technical writing: this alone has increased the enrollment in the course by some four hundred students a year.

We mention this to indicate that colleges teaching technical writing need more teachers. Some of these teachers have already prepared themselves by obtaining special degrees at such schools as Rensselaer, Carnegie-Mellon, and Colorado State. Others are branching out into what is a completely new field for them.

Schools of journalism are now recognizing that their graduates may get jobs in science writing for newspapers and journals. They may also end up in publicity or advertising with a heavy science slant. All of these students of technical journalism must be instructed by qualified teachers. For this reason, openings for technical communication instructors are multiplying.

You will find that most positions require a Master's or a Ph.D. degree, and that administrators favor applicants with some experience in teaching technical writing. The problem may then be how to qualify for these positions, especially if you are coming from a traditional English department.

Some universities are now developing and offering courses, particularly on the graduate level, in the practice and teaching of technical writing. This is usually the result of an English department's awakening to the opportunities open to its students in this area, especially to students who already have some expertise in technical writing. A typical graduate-level course of this kind would offer instruction in business and technical communication, providing you with basic texts, study outlines, and assorted assignments and exercises.

If departmental courses are not available, you have other options. Several universities, among them the University of Michigan, MIT, and Rensselaer, offer week-long institutes and seminars. Here are opportunities to network with many people, both experienced and inexperienced, trade ideas, and get a real feel for this comparatively new discipline. These institutions and seminars regularly cover useful classroom topics, such as types of technical writing courses, designing objectives for technical writing courses, report writing topics and assignments, evaluation and grading of student papers, and computer-assisted instruction. Frequently these programs will include information about resource material available to technical writing teachers, areas of needed research, and consulting possibilities. Most of the programs will include workshops in which the participants practice various technical writing skills to give them a better understanding of some of the problems their students may face. Any technical writing teacher will gain valuable experience and acquire much useful information by attending such a program, and the teacher seeking a technical writing teaching position will enhance his or her credentials by participation. The annual conference of The Society for Technical Communication (STC) is also a valuable source of information. What better way to learn than to be able to talk with professional writers and with well-known teachers. At the STC conference you may also learn about the prevailing job market and meet with potential employers. Of late, associations such as the Modern Language Association and the Popular Culture Association have been including panel sessions on various aspects of teaching and research in technical writing in their national and regional meetings. This is concrete testimony of an increased awareness of the importance of technical writing teaching at a time when there are many cutbacks in other teaching areas.

Once you have gained a position teaching technical writing, you do not need to struggle along unaided, even if you are the only person in your school teaching the subject.

## TEACHER RESOURCES

A basic exposition syllabus, coupled with a reliable technical writing text and any supporting materials gathered from the sources already mentioned, is a starting point. The new teacher can draw on a variety of resources. *The Technical Communication Quarterly,* a journal, provides many useful suggestions contributed by experienced teachers. In the same category is the *Journal of Technical Communication.* Perhaps the most attractive feature of this publication is the mixture of articles by both teachers and industrial writers. You can easily carry over into your classes many of the ideas suggested by practicing writers. Some of these articles have been made available through an anthology, *Opportunities in Technical Communication* by Gould and Losano, listed in our appendix of recommended references.

The STC provides much useful information through its magazine, *Technical Communication.* STC also publishes a series of specialized collections including *Teaching Technical Writing* and *How to Teach Technical Editing.* The National Council of Teachers of English has put out several pamphlets on teaching technical writing, and their journal, *College English,* publishes some fine articles on technical writing which the new or experienced teacher will find useful. A bibliography of references for many areas of technical writing, as well as a list of periodicals and journals, is located at the back of this book in Appendix A.

New teachers of technical writing will often have the benefit of working with an experienced technical writing teacher, who will provide direction and useful materials to support the new teacher. Without such help, a new teacher can survive by relying on his or her background in English composition along with the realization that technical writing is practical in nature (there is no place, for example, for the leisurely essay we may find in a composition text). The new teacher, supported by a good technical writing text, should assign written exercises that reflect the real needs of his or her students. Some brief reading in technical journals or in *Science* or *Scientific American* may be especially useful in making the technical writing class more relevant to students.


## CONSULTING AND FREE-LANCE WRITING

Much has been written lately about the possibilities of consulting in technical writing—going into a company to help it solve its writing problems, perhaps through a series of in-house training courses or through individual work with the company's writers and editors. Other articles discuss possibilities as a free-lance technical writer. *Technical Communication,* for example, has explored consulting in depth.

Those interested in the consulting aspect of technical communication must first consider the amount of competition in the field. Without a reputation or some recognized affiliation, with a university for example, the consultant will get little work. It is difficult for the individual to develop a lucrative consulting business when several large communication consulting firms exist and many highly experienced technical writing instructors offer their services as consultants. Ordinarily we see consulting as an added benefit of college technical writing teaching, but recently we have seen numerous advertisements for consultants to work with national companies. Usually these positions require some teaching background, most require a Ph.D., and all require extensive travel. In such positions, the technical writing consultant would offer intensive short courses for various industries and organizations throughout the country and possibly in foreign countries. This is demanding work, usually high paying, and may represent a career choice for those unable to gain secure university teaching positions.

Free-lance technical writing may offer more possibilities for the newer professional. Often small organizations that cannot afford permanent full-time technical writing personnel will hire others on a temporary basis or will have work available for writers to do at home. For several years, one of our students supplemented her income quite nicely by documenting computer programs at home, using a terminal supplied by the company. Another student gained valuable experience writing and editing a study guide for the Certified Public Accountant exam and another for the LSAT. For this work, around 20 hours a week, she was paid $12.00 an hour. Obviously, more experienced technical writers would charge more. Part of her job was to design reading comprehension tests (for the LSAT). This included selecting relevant passages and making up reading questions that were followed by the correct answers to the questions and explanations of the answers. For students and fledgling technical writers who wish to develop some credentials in the profession, this is a good way to begin; for the experienced professional who prefers on-call work or to work in his or her home on a variety of projects for a variety of clients, free-lance technical writing offers worthwhile possibilities.

# CHAPTER 4

# EDUCATION, TRAINING, AND COSTS

Most technical writers are graduates of four-year programs and have earned Bachelor's degrees. Some have Master's degrees gained through specialized course work beyond their undergraduate study. However, a growing number of technical writers have come out of two-year and community colleges.

We would not want to give you the idea that you can't become a technical writer directly upon high school graduation. But recent reports issued by the Society for Technical Communication show that the level of education for writers has risen considerably over the past two decades.

What constitutes good training? We have selected the views of two professionals who have had enough experience in the field to realize the educational needs of the prospective technical writer. Fred W. Holder, who has written a great deal about communication, expresses his views this way:

> Ideally, a candidate should have a Bachelor's degree in engineering (in the particular specialty with which you're dealing) and a Master's degree in English, journalism, or another field requiring a sound background in written communication. . . . I've found that people with sixty to ninety credit hours of college work covering English, journalism, mathematics through calculus, physics, chemistry, and a wide range of other subjects make excellent technical writers.

Marguerite F. D'Amico, of the Western Electric Company, expresses her views on the subject this way:

> There are three essential requirements for those involved in translating and presenting technical ideas: a solid foundation in the basic sciences and some understanding of how they relate to technology; an understanding of how to organize and present concepts clearly, logically, and graphically; and a sensitivity to the standards and needs of those receiving and supplying the information.

What does it all add up to? It simply means that technical writers will need more and more formal education as time goes on. In discussing technical writing education, some general principles apply:

- There is a differentiation in most companies between technical writing and technical editing. Editing requires a person who is adept at improving the composition end of writing—correcting grammar and punctuation, style, and construction of sentences and paragraphs. Technical writing, on the other hand, encompasses the whole process. It takes in editing, of course, but it extends to original writing as well as the rewriting of other people's manuscripts. The writer must have a firm grasp of the technical material to cope with this kind of assignment.

- For the rather restricted job of technical editing, it is generally agreed that the person who is trained in English composition will do well. A prospective technical editor should also possess, of course, an affinity for technological subjects and familiarity with engineering and scientific terms.

- For the writer who must deal in depth with technical subjects, a firm foundation in science and engineering is essential.

- For both the technical writer and the technical editor, some knowledge of and aptitude for working with computers and word processors of one kind or another are essential. This includes methods of transmitting information with the aid of computers, information storage and retrieval systems, and various reproduction devices.

## COURSES IN TECHNICAL WRITING

Over the years, colleges and other schools have recognized that engineering students not only should be taught English composition, but also should be exposed to courses in technical writing. These courses are usually taught by members of the English department in an engineering college or by teachers of engineering who have an interest in writing. They deal with special forms of technical writing such as report writing and the preparation of scientific papers and magazine articles.

As a result of the formation of various technical writing societies and the great need for technical writers, industry and the technical press have taken more interest in what is being taught in colleges. Every year the Institute of Electrical and Electronics Engineers (IEEE), numbering nearly 200,000 people, holds a special session titled "Engineering Writing and Speech." During this session, seminars and panel discussions on the training of engineers are held to foster clearer and more informative written communications and to improve the relationships between engineers and technical writers. The result of this two-way process has been the introduction of many fine technical writing courses and four-year programs into a number of colleges and universities. A considerable number of schools now offer majors in this discipline. The programs have been given various names and can be found in communication- or humanities-oriented departments under such course titles as: science writing, science information, technical journalism, and technical communications. In addition, departments of business writing in business colleges now offer technical writing, as do some schools of journalism. In the following pages you will find discussions of some of the technical writing programs.

It is important to consider two things: what programs and courses are available, and what to expect when you become employed.

We carried out a study of this very subject among a group of technical writers. These are the answers broken down into three categories:

### WHAT ARE YOUR PRESENT DUTIES?

**Professional**
Preparing computer manuals
Hardware manuals
Reports and proposals
Audiovisuals
Brochures
Layout

**Management**
Writing supervision
Consulting
Production operations
Editorial management
Training programs

**Publicity**
Placing technical articles
Writing technical articles
Preparing brochures
Preparing newsletters

**Academic**
Teaching technical writing
Media instruction
English composition

### WHAT OTHER COURSES IN ADDITION TO TECHNICAL WRITING SHOULD BE INCLUDED IN THE CURRICULUM?

Science or engineering courses
Media courses using audiovisuals and cassettes
Oral presentations

## WHAT COURSES SHOULD BE TAKEN OUTSIDE THE TECHNICAL WRITING FIELD?

Management administration
Sociology
Industrial psychology
Computer science
Graphic arts
Photography
Printing

From this information, you may be able to extract a couple of pointers. First, it would be a good idea to know what kind of technical communication job you are aiming for—writing or editing; in what area you think you are qualified—dealing with reports, manuals, papers and articles, publicity, or advertising. Then, you should find a college that fits your requirements.

The 1993 revision of *Peterson's Guide to Four Year Colleges* lists many colleges in the United States and Canada that offer B.S. degrees in technical writing. Enough information about entrance requirements and approximate cost is included to allow you to decide which schools you wish to contact for further information. It does not provide details about the courses or programs, but includes addresses for requesting information.

To illustrate sample curricula, a number of schools—both universities and community colleges—have been randomly selected and their technical writing programs have been briefly outlined on the following pages. It is beyond the scope of this textbook to provide more detailed information.

### Alderson-Broaddus College

An undergraduate major in humanities with a concentration in technical writing is offered at Alderson-Broaddus. The program prepares writers for business, science, education, and social work. Course work includes technical writing, computer programming, data analysis, and independent study. Also required are at least two terms of practicum experience. Write:

Humanities Department
Alderson-Broaddus College
Philippi, West Virginia 26416

## California State University at Fullerton

At California State University, technical and business communication is one of the seven fields leading to a Bachelor of Arts degree in communication. In addition to preparing students for media careers, the program emphasizes the broad principles of communications, the function of the mass media in a democratic society, and theories relevant to informing, instructing, and persuading through the media. The following courses are required of all communications majors: Mass Communications in Modern Society, Communications and the Law, and History and Philosophy of American Mass Communications. Other courses include Principles of Communications Research, Persuasive Communications, and World Communications Systems. Write:

Communication Department
California State University at Fullerton
Fullerton, California 92634

## Carnegie-Mellon University

Carnegie-Mellon offers a four-year program in technical writing and editing leading to a Bachelor of Science degree. The aim is to equip the student with as broad a base as possible for a professional career in technical communication. These are the course requirements and electives: four courses in humanities, two courses in the social sciences, four courses in literature; courses in technical research and report writing, layout and design, and graphic arts; and two courses each in mathematics, chemistry, and physics. Students serve a short internship in industry. Write:

English Department
Carnegie-Mellon University
Pittsburgh, Pennsylvania 15213

## Clarkson College of Technology

The technical writing program at Clarkson consists of a major offered within the Department of Humanities, leading to a B.S. degree in Technical Communication. It is offered to students interested in combining the study of science, engineering, or management with a study of communication, publication, and mass media.

Upon graduation, students pursue such careers as technical writing and editing, technical journalism, computer documentation, advertising writing and design, and publication management.

Among the ten courses making up the program, the following are required: Theory of Rhetoric for Business, Science, and Engineering; Business and Industrial Report Writing; Engineering and Scientific Report Writing; Theory and Philosophy of Communication; Business and Professional Speaking; and Technical Communications Internships or Research Projects. Write:

Department of Humanities
Clarkson College of Technology
Potsdam, New York 13676

## Colorado State University

The technical writing curriculum at Colorado State is part of the Department of Technical Journalism. According to a course prospectus, technical journalism is the "seeking, interpreting and reporting of information for mass and specialized audiences." CSU journalism students may prepare for jobs with business, industrial, scientific, and popular magazines, and for work in radio, television, documentary films, newspapers, photojournalism, public relations, and technical writing for public and private organizations.

In addition to journalism skills, students in this program gain a broad background and in-depth exposure to a variety of social and natural sciences, the humanities, business, and other elective subjects. The Department of Technical Journalism carries on cooperative programs with the journalism profession. Elements of these programs are internships, practicums, and class projects with practicing professionals. Write:

Department of Technical Journalism
Social Science Building
Colorado State University
Ft. Collins, Colorado 80523

## Massachusetts Institute of Technology

Four distinct channels of instruction in technical communication leading to a Bachelor of Science degree have been designed for the MIT program. These are elective undergraduate and graduate courses in technical writing and science writing; instruction in technical communication in conjunction with an undergraduate engineering or science course; a concentration of writing courses as part of a B.S. degree in writing and literature; and a continuing education summer program on communicating technical information.

Course requirements and electives are in science and mathematics, plus eight humanities subjects, including six writing courses and literature. Write:

Writing Program
Department of Humanities
Massachusetts Institute of Technology
Cambridge, Massachusetts 02139

## Metropolitan State College

The Communications Multimajor Program at Metropolitan State leads to a Bachelor of Arts degree. The program enables any department to design its own communication system based on its goals, staff, and resources. Core courses are Introduction to Communication Theories and Psychology of Communications or Techniques of Persuasion. The Communications Multimajor Program has developed these tracks: Visual Communications, Industrial/Organizational Communication, and Cooperative Program for Careers in Communications. Write:

School of Liberal Arts
Metropolitan State College
1006 11th Street
Denver, Colorado 80204

## Michigan Technological University

The MTU program for a Bachelor of Arts in Scientific and Technical Communication requires forty-five hours in the core option, with courses in writing, editing, film and television, and the history and theory of communications. To this option must be added forty-five hours in an area of science and/or technology.

The technological courses are selected from the life sciences, basic science, general engineering, forestry and conservation, and other appropriate areas. Write:

Department of Humanities
Michigan Technological University
Houghton, Michigan 49931

## Montana College of Mineral Science and Technology

Professional and technical communications is one of three options open for a Bachelor's degree under the general title of society and technology. The other two options are technology and public policy, and human values and technology.

The concern of the first option is to train students in written, verbal, and graphic arts in a framework of scientific and engineering problems.

These are some of the principal courses in the communications curriculum: Introduction to Journalism, Mass Media, Elementary Photography, Technical Editing, Printing Processes, Industrial Graphics, and Interpersonal Communication. Write:

Department of Humanities and Social Sciences
Montana College of Mineral Science and Technology
Butte, Montana 59701

## Oregon State University

The technical journalism program at OSU combines communication skills and course work in scientific and technical disciplines. In addition to completing twenty-eight to forty hours in journalism, a student must complete one of thirty-one technical minors. The scientific-technical minors include animal science, business administration, fisheries and wildlife, food science and technology, forest management, aerospace studies, computer science, and earth sciences. Write:

Department of Journalism
Oregon State University
Agriculture Hall 229
Corvallis, Oregon 97331

## Rock Valley Community College

The purpose of the two-year program in technical writing is to provide job entry and job improvement education. The curriculum has been prepared in answer to the needs of local industry. The technical writing curriculum includes a minimum of twenty semester hours in composition and twenty hours of concentration in a technological area, such as electronics, aviation, or industrial engineering. Technical writing internships provide on-the-job experience in a variety of report writing areas. The program leads to an Associate of Applied Science degree in technical writing. Write:

Rock Valley Community College
3301 North Mulford Road
Rockford, Illinois 61101

### South Dakota State University

Under the general title of journalism, South Dakota offers a Bachelor of Science degree in science and technical writing. This is a major for those who wish to become technical writers, either for commercial companies or for magazines and newspapers. By the beginning of the junior year, the students should decide whether to emphasize the physical sciences, the biological sciences, or technology, and they must select an additional twenty credits in science or technology. Sequences are available in news-editorial, advertising and broadcast journalism, agricultural journalism, printing-journalism, and printing management. Courses in mass communication are also given. Write:

Department of Journalism
South Dakota State University
Brookings, South Dakota 57006

### University of Minnesota

The major in technical communication at the University of Minnesota leads to a Bachelor of Science degree. It has been designed to develop professional communicators trained for professional opportunities or to serve as a base for graduate studies in the discipline.

These are some of the requirements and electives for the University of Minnesota program: Communication, Language, and Symbolic Systems; Artistic Expression: Technical Communication, including writing and editing, media communication, graphics, organizational, managerial, and training communication, theory and research; and Technical Electives. Write:

Department of Rhetoric
University of Minnesota
St. Paul, Minnesota 55108

## University of South Dakota at Vermillion

The university offers a technical communication program. An option elected by most of the students is to complete a Bachelor of Science degree in a specialized area and also complete the Associate of Applied Science degree in technical communications. The aim of this program is to develop a technical communicator who is able to interpret scientific and technical language so that it can be used by management specialists, industrial technicians, and consumers. Write:

Director of Admissions
University of South Dakota at Vermillion
Vermillion, South Dakota 57069

## University of Washington

This university offers two routes by which one can prepare for a career in scientific and technical communication. Both combine a solid grounding in science and mathematics with training for writers or editors in technically oriented organizations.

In the college of engineering, the program requires concentration in scientific and technical communication, electives in related aspects of communication, and an actual technical writing and editing project.

Through the College of Arts and Sciences, students may obtain a different background, particularly in science and general studies. For example, students wishing to enter a career in medical writing could combine courses in the biological sciences and medicine with communication courses. Write:

Department of Humanistic Social Studies
356 Loew Hall FH-40
College of Engineering
University of Washington
Seattle, Washington 98195

### University of Waterloo

In Canada, the University of Waterloo in Ontario offers a Bachelor of Science degree in technical writing. Write:
University of Waterloo
Waterloo, ON N2L 3G1

### University of Wisconsin-Stout

This university offers a Bachelor of Science degree in industrial technology and technical communication. Courses are required from the graphic arts, speech, and learning resources departments in addition to courses in industrial management. The general requirements for the degree include a broad background in science and mathematics and a basic core in the liberal arts. In the technical writing option, the student can choose among courses in advanced technical writing, journalism, and rhetoric. Write:
Industrial Technology Department
University of Wisconsin-Stout
Menomonie, Wisconsin 54751

## GRADUATE STUDIES

The following colleges and universities offer graduate programs in technical and scientific communication.

## Boston University

The Master of Science degree at Boston University is given at the completion of a major in Science Communication. Students are required to take work during three semesters and a summer session.

A statement by Boston University describes the program as providing students with journalistic skills to apply to the communication of scientific, engineering, and technological information. Graduates qualify for positions as reporters; writers and editors for technical, scientific, and business magazines; writers and editors for publishing houses; and communication specialists and administrators.

Some of the courses are: Basics of Science Journalism; Science, Technology, and Public Policy; Instructional Video Tape Production; and Medical and Demographic Reporting. Write:

School of Communication
Boston University
640 Commonwealth Avenue
Boston, Massachusetts 02215

## Illinois Institute of Technology

IIT's graduate program in Science Information is an interdisciplinary curriculum for the management and processing of technical and scientific information. Graduates are able to service the information needs of research and development organizations, government agencies, academic institutions, and libraries. Course requirements lie in four areas: information collection, storage, and retrieval; writing, editing, abstracting, and interpreting information; management of information centers; and the utilization of library and other information sources. Write:

Director of Science Information
Life Science Building
Illinois Institute of Technology
Chicago, Illinois 60616

## Miami University

Miami University has a program leading to the Master of Technical and Scientific Communication (M.T.S.C.). The M.T.S.C. program is professional and practice-oriented to prepare students who specialize in technical, scientific, and other fields to communicate their knowledge in an understandable and usable way. For instance, graduates of the program might write, edit, or supervise the creation of instruction manuals, grant proposals, scientific research reports, or slide-tape shows on technical subjects, to mention just a few of the possibilities.

The interdisciplinary program consists of eight required courses plus three electives. Also required is a one-semester internship in which students work as apprentice technical and scientific communicators in business and government. Students who are already working in the profession may perform the internship with their present employers; those with substantial professional experience may choose to write a thesis in lieu of the internship. Students in the program will prepare many of their assignments in the Technical and Scientific Communication Laboratory.

Miami University also has a program, "Writing for Business, Industry, and Government," for those who want to receive a Bachelor of Science degree in English and pursue a career in writing and editing. Write:

Department of English
356 Bachelor Hall
Miami University
Oxford, Ohio 45056

## Rensselaer Polytechnic Institute

The Master of Science program in technical writing and communication seeks to develop communication specialists and to prepare students for further advanced study. Students enter this program with academic backgrounds ranging from education to engineering. Those with undergraduate degrees in the humanities are able to broaden their backgrounds in preparation for a wider variety of professional careers; those with technical or scientific undergraduate degrees are able to polish their communication skills and to integrate their undergraduate studies into the field of communication theory.

No matter what their academic backgrounds, students are required to take thirty credit hours beyond the Bachelor's degree and can complete the course in one year of full-time study or several years of part-time study. Some of the courses included in this program are Visual Communication, Film and Fiction; Language: The Cultural Milieu; Data Processing; Organizational Psychology; and Advertising Strategies and Promotion.

A Bachelor of Science degree in communication is also offered at Rensselaer. The program consists of four specializations or options: technical writing, communication theory and research, communicating arts, and literature. Write:

> Department of Language, Literature, and Communication
> Rensselaer Polytechnic Institute
> Troy, New York 12181

## OTHER PROGRAMS

A number of colleges offer courses that give excellent training in professional writing but do not require exclusive specialization in this field:

- *Air Force Institute of Technology,* Wright-Patterson Air Force Base, Ohio. Several courses in technical communications have been developed: Technical Writing, including the analysis of the communication situation, selecting, organizing, and presenting technical information, and style and format; Technical Reports and Theses, in which oral reports are covered; Technology and Management; and Seminars in Technical Communication, with practice in writing technical papers and research projects.

- *Arizona State University,* Tempe, Arizona, offers technical writing courses in description, research methods, and professional writing.
- *Baylor University,* Waco, Texas, offers technical composition.
- *Bowling Green State University,* Bowling Green, Ohio. A program in technical communications is being offered, including two courses in technical writing for undergraduate and graduate students.
- *Brigham Young University,* Provo, Utah. A composition and reading course has been designed to develop accuracy and skill in writing scientific pamphlets, articles, and reports.
- *California Institute of Technology,* Pasadena, California, gives a three-credit technical writing course.
- *California State Polytechnic College,* San Luis Obispo, California, offers a major in technical journalism, including such courses as photojournalism, introductory journalism, editing, editorial and feature writing, layout, and applied technical journalism.
- *Central Michigan University,* Mount Pleasant, Michigan, teaches technical and professional writing, including conducting research, organizing data, and reports.
- *Clarion State College,* Clarion, Pennsylvania, has established an undergraduate program in communication. The program has been developed for students interested in business, industry, and government agencies.
- *Colorado School of Mines,* Golden, Colorado, teaches advanced technical exposition. Among the items reviewed are reports required by industry and government and a survey of the laws and guidelines for preparing environmental impact statements.

- *C. W. Post College, Long Island University,* Brookville, New York, gives instruction and practice in the preparation of reports, documents, and industrial correspondence.
- *Erie County Technical Institute,* Buffalo, New York, has a communications course that includes the writing of reports, abstracts, and technical business letters.
- *Fairmont State College,* Fairmont, West Virginia, has an undergraduate course in formal and informal reports and business correspondence.
- *Fordham University,* New York, New York, through its adult education department, offers courses in technical writing.
- *Georgia Institute of Technology,* Atlanta, Georgia, gives courses in technical reports and letters.
- *Hartwick College,* Oneonta, New York, in its course titled Business Writing, gives practice in writing memos, letters, and brief reports as solutions to communications problems.
- *Hinds Junior College,* Raymond, Mississippi, places its emphasis in technical writing on the practical application of writing skills. The technical writing curriculum contributes to an Associate of Applied Science degree.
- *Hudson Valley Community College,* Troy, New York, teaches report and letter writing for business and industrial personnel, with emphasis on format, organization, and style.
- *Husson College,* Bangor, Maine, offers instruction in research and report writing, various forms of business writing, and letters and reports.
- *Indiana University at South Bend,* Indiana, offers the course Professional Writing Skills. It includes technical reports, proposals, and papers.
- *Iowa State University,* Ames, Iowa, offers a curriculum in agricultural journalism, which includes specialized technical writing courses, such as technical reporting, copy editing and typography, technical advertising, and journalism.

- *Marietta College,* Marietta, Ohio, offers an undergraduate course in business and technical report writing and in the techniques of collecting and organizing material.
- *Merrimack College,* North Andover, Massachusetts, gives instruction in the exploratory forms commonly used in science, business, and research. The course has been designed to enhance technical writing skills.
- *Midwestern University,* Wichita Falls, Texas, gives a course featuring abstracts, technical description, and report writing.
- *Mohawk Valley Community College,* Utica, New York, provides training in the preparation of reports. Students are required to write several short reports in both technical and general education areas.
- *Montana State University,* Bozeman, Montana, gives a career-oriented program of communication, including studies in media, technical reporting, proposal writing, and graphics.
- *New Jersey Institute of Technology,* Newark, New Jersey, specializes in engineering report writing, industrial reports, technical publications, and editing. Business writing and oral presentations are also taught.
- *North Carolina State College of Agriculture and Engineering,* Raleigh, North Carolina, offers a course in technical writing.
- *Northeastern University,* Boston, Massachusetts. Several courses in technical writing are given, including the history and the principles of technical writing, proposal writing, electronic technical manual writing, and computer software technical writing.
- *Northern Michigan University,* Marquette, Michigan, provides scientific, technical, and report writing, in which formal expository writing and the preparation of reports are presented.
- *Northwestern University,* Evanston, Illinois, includes technical writing in its graduate school of journalism.

- *Oklahoma State University,* Stillwater, Oklahoma. A technical journalism option is combined with agriculture, home economics, and technical report writing.
- *Pennsylvania State University,* University Park, Pennsylvania, through its courses—Technical Writing, Business Writing, and Advanced Technical Writing and Editing—provides an option leading to a Technical Writing Certificate.
- *Polytechnic Institute of New York,* Brooklyn, New York, offers course work in basic report writing and communications, from technical correspondence to professional articles.
- *Purdue University,* West Lafayette, Indiana, schedules a sequence in journalism and journalistic applications.
- *State University of New York Agricultural and Technical College,* Farmingdale, New York, provides studies in written and oral reports, proposals, resumes, and letters of application.
- *Texas A&M University,* College Station, Texas, provides an undergraduate writing specialization consisting of four courses in English: audience analysis, technical writing, editing in business and industry, and the writing of technical speeches and other presentations. A graduate course in technical editing is planned for the future.
- *Texas State Technical Institute,* Waco, Texas, has a six-quarter, technical communication program to prepare entry-level technical communicators. Courses emphasize rhetoric, reports, manuals, technical journalism, and computer science.
- *Tulane University,* New Orleans, Louisiana, teaches courses in written and oral reports.
- *U.S. Air Force Academy,* Colorado Springs, Colorado, teaches technical writing, including the preparation of graphs, tables, drawings, and documentation.
- *University of Alaska,* Fairbanks, Alaska, has listed a number of technical writing courses.

- *University of Arizona,* Tucson, Arizona, offers the analysis and presentation of scientific and technical information.
- *University of Cincinnati,* Cincinnati, Ohio, offers report writing, research methods, interpretation and reorganization of data on the undergraduate level.
- *University of Colorado,* Boulder, Colorado offers an elective in technical papers and article and report writing.
- *University of Florida,* Gainesville, Florida, teaches courses in technical writing, business communication, and proposal preparation.
- *University of Idaho,* Moscow, Idaho, offers several technical writing courses.
- *University of Kansas,* Lawrence, Kansas, places emphasis on good expository writing, with practice in specific forms of technical writing.
- *University of Kentucky,* Lexington, Kentucky, offers a communications sequence, including courses in communication process and advertising for mass media.
- *University of Maine,* Orono, Maine, teaches various types of business, professional, and technical writing.
- *University of Maryland,* College Park Maryland, has a program in industrial journalism with courses in industrial communications and the production of company periodicals.
- *University of Massachusetts,* Amherst, Massachusetts, teaches factual exposition, with courses in research methods, bibliography in scientific areas, military specifications, and reports.
- *University of Michigan,* Ann Arbor, Michigan, has developed pioneer teachers in technical communications. An option in technical and professional communications is included in the undergraduate interdisciplinary engineering program.
- *University of Nebraska at Lincoln,* Nebraska, gives a technical writing course that takes the student through note-taking techniques, short papers, and outlining, resulting in a paper.

- *University of Oregon,* Eugene, Oregon, offers scientific and technical writing, with instruction in reports, proposals, and correspondence. It also has courses in business communication, with practice in writing forms pertinent to business, industry, and the professions.
- *University of Pittsburgh,* Pittsburgh, Pennsylvania, gives technical writing courses for undergraduates in the School of Engineering and Mines.
- *University of Tennessee at Knoxville,* Tennessee, offers a technical writing course.
- *University of Texas,* Austin, Texas, offers undergraduate courses, titled Writing for Engineers, Writing for Science Majors, Advanced Technical Writing, and Editing.
- *University of Toledo,* Toledo, Ohio, offers a course in technical writing under the sponsorship of the General Engineering Department.
- *University of Wisconsin,* Madison, Wisconsin, has courses in written and oral report preparation and presentation for undergraduates in the College of Engineering.
- *Washington State University at Pullman,* Washington, has regularly scheduled courses in technical writing.
- *Western State College of Colorado,* Gunnison, Colorado, provides an introduction to technical writing and multimedia competence.

In Canada, in addition to the University of Waterloo's Bachelor of Science degree, at least three colleges offer technical writing courses: Ontario Agricultural College, Guelph, Ontario; Southern Alberta Institute of Technology at Calgary; and Queen's University at Kingston, Ontario.

## HOW TO APPLY

If you are interested in taking a course in technical writing at a college or technical institute, you must follow certain procedures:

- Contact the admissions office of the school to be sure the course is available. Ask for a catalog or request that the admissions office direct you to the department offering the course for further information.
- Examine the catalog to determine if the course is given during the day, in the evening, in the summer, by correspondence, or by any other special arrangement.
- Determine if college credit is given for the course. It is wise to accrue credits; they will be useful if you decide to pursue a degree.
- Find out if a prerequisite is required. This may be in the form of other writing courses, such as English composition. Also find out if you need a high school diploma in order to take the course, and whether or not the course is given on the graduate level and requires a Bachelor's degree for enrollment.
- Find out if you may enroll as a special student, taking only one course, or if you must complete enrollment as a regular, full-time student.

## IN-COMPANY TRAINING

Although the shortest path to becoming a technical writer is by mapping out a definite educational program, there is still another way, and that is through in-company training. Such programs will benefit you if you are an employee of the company and have had formal training but need to get actual writing experience.

The University of Massachusetts at Amherst, through its technical communication teaching program, has done a lot of work with industry personnel and has organized many technical writing workshops.

In-company training programs for technical writers include such practices as giving newly hired persons the company style manual and telling them to work out their projects. Some companies offer elaborate seminars to employees which are taught by professional educators. Many of the nation's most progressive companies, large and small, offer such courses because they find that training programs are an effective way of keeping their people alert and up-to-date. They are also a means of attracting good personnel to the company.

Technical writing training programs fit into several categories. One of these is the formal course offered at regular intervals on company time and run by a company employee who is or has been a technical writer.

Another form of teaching, which is usually conducted less formally, is provided by the company bringing in outside consultants. They are likely to be teachers of technical writing, report writing, and technical composition, teachers well known for their practical experience in industry itself. Such a consultant is Robert R. Rathbone, who has been in charge of the technical writing program at the Massachusetts Institute of Technology for many years. Professor Rathbone purveys his knowledge through lectures, discussions, workshops, outside assignments, and editorial conferences.

## SHORT-TERM COURSES

An outgrowth of education in the technical writing field has been a number of special institutes, seminars, and workshops. They provide short-term means of bringing technical writers up to date on current practices. For the most part, these institutes, especially those run by private institutions, appeal to already established writers. However, they are also valuable to new writers who want to find out what technical writing is all about, meet other writers, and make useful contacts.

The oldest of these special training groups is the Technical Writers' Institute of Rensselaer Polytechnic Institute. This pioneer organization was established with the aim of providing a forum where technical writers from industry could meet to discuss and practice writing techniques under the supervision of experienced teachers in the technical writing field. It conducts teaching sessions, lectures, discussions, and demonstrations by recognized authorities from industry, government, and various technical writing agencies. This particular institute is aimed at: technical editors, writers, and special service groups; engineers and scientists in positions where an ability to write well is essential; and administrators supervising technical writing, training programs, and promotional work.

For further information, write:
Technical Writers' Institute
Rensselaer Polytechnic Institute
Troy, New York 12181

Rensselaer also offers an annual Technical Writing Institute for Teachers. This program, conducted in June, is designed to assist teachers in acquiring the knowledge and skills necessary to teach and develop courses in technical and professional writing at all educational levels. Representative sessions have included: From Teacher of Literature to Teacher of Technical Writing, Bibliography for Technical Writing, Audience Analysis, and Teaching Technical Writing in Industry.

For information, write:
Technical Writing Institute for Teachers
Rensselaer Polytechnic Institute
Troy, New York 12181

The University of Michigan has developed two short seminars in technical writing. One of these is Written Communications for Engineers, Scientists, and Technical Writers, held in August. This course is designed to improve the clarity and efficiency of scientific and technical communications. It focuses on organization, exposition, and style with special problems faced by the scientific and technical author.

The other seminar is Teaching Technical and Professional Communication, held in July. This conference has been designed to improve instruction in technical and professional communications in community colleges, universities, professional schools, industry, and government. Information about these two seminars may be obtained from:

Department of Humanities
College of Engineering
University of Michigan
Ann Arbor, Michigan 48104

At Oklahoma State University, a workshop on Teaching Technical Communication has been established. This workshop has been organized for teachers in secondary schools, vocational schools, and both two- and four-year colleges. Information may be obtained from:

Technical Writing Program
Oklahoma State University
Stillwater, Oklahoma 74074

During one week in October, a short course, Professional Technical Writing Workshop, is offered by the University of California at Los Angeles. This program is designed for executives, managers, physicians, engineers, technologists, and others who are having difficulty with developing and producing technical reports, memoranda, letters, and presentations. The intensive writing assignments and group discussions that follow periodic lectures are presented to illustrate new methods of organizing written materials. Participants learn to resolve their immediate problems and leave with valuable techniques which they can use in the future. Send inquiries to:

UCLA Extension
10995 Le Conte Avenue
Los Angeles, California 90024

A vigorous, one-week program in Communicating Technical Information is held in August at the Massachusetts Institute of Technology. This program has been designed to help engineers, scientists, technical writers and editors, publications managers, and teachers of technical writing. In addition to informal lectures on the principles that underlie successful technical communication, the format provides a forum for discussion and debate and an opportunity for staff consultation on individual writing, editing, and management problems.

At this institute a limited number of scholarships are available to help defray tuition costs. Registration forms may be obtained from:

> Director of the Summer Session
> Room E19-356
> Massachusetts Institute of Technology
> Cambridge, Massachusetts 02139

Rice University annually sponsors a workshop on Teaching Technical and Professional Communication, designed especially for teachers new to teaching technical and professional communication and for administrators responsible for developing new programs. To register, write:

> Special Programs
> Rice University
> P.O. Box 1892
> Houston, Texas 77001

Another workshop, the Practice and Teaching of Technical and Professional Writing, has become a regular feature of Old Dominion University. Usually held in July, it includes such sessions as: Practical Criteria for Designing a Writing Course, the Rhetorical Background in Business and Professional Writing, Consulting, Word Processing, and Creating a Bibliography for Technical Writing Instructors. Information about this workshop can be obtained from:

Department of English
Old Dominion University
Norfolk, Virginia 25501

An annual event is the Institute in Technical Communication sponsored by the Southwestern Conference on the Two-Year College. More information may be obtained from:

Department of English
Hinds Junior College
Raymond, Mississippi 39154

A series of Communication Workshops is offered by Colorado State University. The workshops are for individuals from science, industry, government, and business. The latest workshop included writing and editing, producing 35mm slide presentations, videotape editing and production, public relations, and technical communication. Obtain information from:

Department of Technical Journalism
Colorado State University
Fort Collins, Colorado 80523

Teaching Technical and Professional Writing, a working conference, has been sponsored by the University of Washington for some time. These topics have been featured: Approaches to Technical Writing, Course Design, and Style and Tone. The seminar offers an opportunity to work with nationally recognized teachers of technical writing. It has been designed to help practicing communicators and publication managers, as well as teachers. Write:

Scientific and Technical Communication Program
14 Loew Hall, FH-10
University of Washington
Seattle, Washington 98195

An Institute in Technical Communication is held during the summer at the University of Southern Mississippi's Gulf Park campus. It is sponsored by the Southeastern Conference on the

Two-Year College. Representative topics include Teaching Audience Analysis, Technical Style, Computer Instruction, and Classroom Methods That Work. For information, write:

Department of English
University of Mississippi
Hattiesburg, Mississippi 39401

We would like to turn to the comments of technical communicators themselves as we conclude this section on courses and programs. These comments have been gathered from several studies on the subject:

> You just can't put writers in slots any more; they are doing all kinds of things.
>
> Another reason for the greater use of nontechnically trained writers is that most liberal arts students come out of school with a better knowledge of technology than they did, say, thirty years ago.
>
> Most companies would like to see the graduates equipped with communication skills—and these include more than a nodding acquaintance with graphics and some knowledge of computer techniques.
>
> The company would like to think that its new technical writing employee would be able to move into management and administration.
>
> The technical editor and the technical writer should be in contact with the entire communication process. This is easier in a small company where one or two writers must take on a variety of communication jobs.

## WORDS OF ADVICE

As a future technical writer, whether you are just starting your training or are planning to enter from another field, you should think in these terms:

- See that you have enough science courses compatible with the area of technical writing in which you are interested— such as chemistry, physics, electronics, mathematics, computer science, or engineering.
- If you are still in high school, take all available writing and composition courses. Although courses in creative writing are fine for some forms of professional writing, it is important that your curriculum include courses in science and technology. If you are already familiar with the subject matter of science and engineering and join a company as a technical writer, you have an advantage over the young man or woman who is not trained in science.
- If you have already graduated from high school and cannot plan on the four years required to earn a Bachelor's degree, see what courses are offered by two-year community colleges in your area.
- It may be that you cannot find a definite technical writing program in your own college. In this case, you could make up your own technical writing program by majoring in a science and taking writing courses and elective subjects such as mathematics, economics, and statistics.
- If you have already graduated from college, consider the Master's degree programs at various institutions. The plan you follow may be similar to Rensselaer's graduate program in technical writing, or it may be a program that you design yourself with a technical writing career in mind.
- Think also in terms of the various journalism programs available that have options in technical writing. They will give you good training for a career in the technical and scientific press.
- Gaps in your training can be filled in by extension courses, both day and evening, and by correspondence programs.

A word of caution, however, should be added here in connection with correspondence courses. You should be sure that the school offering the course is well established, that it is properly licensed, and that it operates under proper state and federal regulations. A potential employer must have confidence in the sources from which you have received your education, so it is worthwhile to check the accreditation, reputation, and longevity of any correspondence school you are considering. Ask your career counselor or other school administrator for guidance.

## EDUCATIONAL COSTS

The cost of obtaining a degree in technical communication, as with other professions, depends on the kind of school you attend and where it is located. Liberal arts colleges tend to have lower tuition; colleges of engineering and science, higher tuition. State universities are less expensive than private schools. College tuition and costs for room and board keep rising every year. The following figures have been arrived at by averaging costs from a considerable number of colleges.

### State University

If you attend a public university within your own state, the annual costs may be in this range:

| Costs | Low | High |
|---|---|---|
| Tuition | $4,500 | $7,500 |
| Activity fee | 200 | 400 |
| Room and board | 2,500 | 4,800 |
| Health insurance | 200 | 400 |
| Books and supplies | 500 | 800 |
| TOTAL ANNUAL EXPENSES | $7,900 | $13,900 |

## Community College

The fees for community colleges vary from state to state and are adjusted according to the place of residence of the student—within the county or district, outside the county, or outside the state.

## Private College

| Costs | Low | High |
|---|---|---|
| Tuition | $6,300 | $17,000 |
| Activity fee | 200 | 400 |
| Room and board | 2,500 | 6,500 |
| Health insurance | 200 | 400 |
| Books and supplies | 500 | 800 |
| TOTAL ANNUAL EXPENSES | $9,700 | $25,100 |

## Graduate Education

Graduate education is run somewhat differently from undergraduate. The principal difference lies in the fact that a great many students can afford to go to graduate school only with financial aid—tuition scholarships and fellowships offered by colleges and industries. The scholarship provides tuition only; the fellowship usually contains a modest living allowance as well as the stipend for tuition. In addition, there are assistantships whereby the graduate student is assigned to a particular department—teaching undergraduate classes, correcting papers, or assisting in laboratories. In order to fulfill these assignments, the graduate student is not permitted to take a full academic load. This means that it will take longer than two or three semesters to complete a master's program and may take twice as long. Under these circumstances, it is impossible to know in advance exactly how much the program will cost.

Costs of education vary from year to year and are affected by the amount of financial aid available in the form of state scholarships, fellowships, and assistantships. In addition, part-time work in writing and editing may be available on the campus, especially to technical writing and journalism students.

## SCHOLARSHIPS AND FELLOWSHIPS

You will find a variety of financial aid sources available at colleges and universities. The following is not a complete list, but it will give you some idea of where to apply for assistance.

*Society for Technical Communication.* The Society awards a number of scholarships annually to students in both undergraduate and graduate schools to carry on specialized study in technical communication. The STC scholarships are administered by those university and technical institutions with degree programs. Recent awards were given to a senior at the University of Washington, at that time majoring in scientific and technical communication; a sophomore at Michigan Technical University, also majoring in scientific and technical communication; and a Ph.D. candidate in communication and rhetoric at Rensselaer Polytechnic Institute. A number of local STC chapters also have scholarships available. Inquiries about these awards should be sent to the Scholarship Committee, STC, 901 N. Stuart Street, Suite 304, Arlington, VA 22203.

*Naval Underwater Systems Center.* This government center in Newport, Rhode Island, offers summer scholarships for college graduates enrolled in or planning to enter a graduate program in technical writing. Salaries for the internships vary, depending upon grade point average, class standing, and availability of funds. During the internships, students become familiar with the Navy laboratory system and with the Center; receive a detailed orientation to the Technical Information Department, including reports, manuals, and graphic arts; and are assigned such projects as editing reports, revising manuals, writing brochures, conducting literature surveys, and developing film scripts.

*Office of Cancer Communications.* This group has established a graduate student internship at the National Cancer Institute in Bethesda, Maryland. The internship provides valuable on-the-job experience in three areas: medical-science writing, health education, and general biomedical communication. Participants have the opportunity to prepare reports on advances in cancer research and to work in such other formats as feature articles, news releases, press summaries, and audiovisual presentations.

Other assistance programs are available. At Texas A&M University, a number of employers in Texas have cooperated with the university by entering an in-company writing program. The Department of Technical Journalism at Colorado State University cooperates with professional firms and practitioners to ensure that as many of its qualified students as possible receive on-the-job training and experience. The student is paid a nominal salary and works within a prescribed area of mass communication. Rensselaer Polytechnic Institute offers assistance through its co-op program with industry.

In most schools, scholarships and fellowships in technical communication are part of the general assistance policy. If you are applying to any of the colleges with degree programs, be sure to inquire at the same time about financial aid since scholarship awards and assistance programs are constantly changing.

# CHAPTER 5

# EMPLOYMENT OUTLOOK

At this writing, the employment outlook for technical writers continues to be good. The strength in the employment field has developed because of several significant factors in technical communication.

One of these is that the well-trained writer is in the forefront of new techniques. If you are a student, you must prepare yourself by getting a solid background in the sciences—particularly in computer science. If you graduate with this competence, you should be able to fill the needs of industrial and government employers.

Another factor is that the well-trained writer can bridge the gap between technical subjects and nontechnical readers. You can bank on one basic fact—general readers must learn something about technology. It is a major facet of modern life. Ultimately, the general public buys the products and services that technology provides. The technical writer informs and persuades laymen through reports, manuals, news releases, articles, and advertising. Thus the technical communicator remains in demand to produce "reader-friendly" articles and instructions which can be easily understood and followed by the nontechnical person.

Don Bissell, publications manager for Sundstrand Corporation, is responsible for audiovisual preparation, writing services, technical publications, and writer training and development. In a recent speech, Mr. Bissell said:

> Need I speculate that we'll be needed even more during
> the next ten years? . . . We have made significant advances
> in the last several years. We [technical communicators] are
> not solidly in control of the next ten years unless we
> recognize and keep pace with a burgeoning technology.

A young woman from the personnel department of an electronics company echoes this statement by stating that the increased interest in technical writers can be attributed to a growth in the use of computers by relatively unsophisticated employers. Computers are now handy tools in many small businesses. While large companies can afford to employ computer specialists, in small companies the employees themselves must become the specialists with the aid of properly written instructions.

This, then, is another possible employment situation to explore. If you have any interest in data processing or computer science, try the smaller companies—and make sure that you include computers in your technical writing curriculum.

You should consider employment opportunities long before you graduate from college. Technical writing is a highly specialized profession requiring a combination of technical training and competence in communication.

In discussing the employment outlook for technical writers, it is valuable to emphasize the following points:

- Some specializations seem to offer greater opportunities than others. Publications departments engaged in producing manuals and instruction books show the greatest expansion and turnover of personnel.
- Research is the starting point of manufacturing; therefore, there will always be a place for technical writers in public and private research institutes. Research reports of all kinds must be written to provide vital information for product development and manufacturing.

- You will find that when you apply for a job, the greatest response will come from employers needing technical writers who specialize in either manuals or research reports.

The technical press—magazines, journals, and publishing houses—employs fewer writers. However, interest in this field has increased, and it is to the credit of publishing firms that they are realizing that second-best writing on technical subjects is not enough; that in a highly competitive field, they must employ technically trained writers. Concerning technical journalism, Victor J. Danilov, executive editor, *Industrial Research Magazine,* has analyzed the situation this way:

> More and more journalistic jobs are open to engineers and scientists. Some of the more common opportunities include: Science reporting for newspapers; professional and trade journal writing and editing; technical and industrial publicity work; science writing for radio and television; and freelance technical writing.

The growing number of openings in technical journalism is a reflection of the increased interest in engineering and scientific news at both the lay and technical levels. An interesting opportunity has arisen in some of the businesses that support large technical firms. Advertisers are just beginning to realize their full potential in the technical advertising field—and with this comes the realization that the technical writer is almost indispensable. The same thing is true of technical publicity. Advertising agencies are trying to locate technically trained writers or engineers with a flair for writing.

We would like to complete this section by emphasizing a very important point. Employers judge prospective employees by their training and education. But the best training in the world may not get you the job if you are lacking certain personality traits. Technical writers are people, not machines; they must work with other people.

Kenneth W. Chew, manager of publications and illustrations at the Rocketdyne Division of Rockwell International, puts great stress on technical writers and their interplay with other people in the company:

> Tact and diplomacy are so important to the writer-editor that too much cannot be said of them. When preparing an original manuscript, the writer must establish and maintain open lines of communication between himself and the source of the material. . . . In an editorial capacity, the writer-editor must rely on his power of friendly persuasion. . . . Initiative and an inquisitive nature are as important as a keen well-developed sense of order. . . . The very nature of communication forces the writer-editor to work at once independently and jointly.

Most interviewers can accurately judge the personality traits of people who will be readily accepted by their fellow workers.

## SOME SUCCESS STORIES

The history of technical communication contains many stories of young men and women who are finding success in the profession. Most of them have broadened their scope and demonstrated that the communications field is a land of opportunity. We have made a survey of several hundred people. Now we pass our findings on to you to demonstrate the variety of their occupations and something of their education.

One young man received B.A. and M.A. degrees from a leading university. In one company, he rose to a position in which he managed a thirty-person staff in four service groups: technical writing, advertising, public relations, and library administration. He then transferred to a research company specializing in optics and computer technology where he reorganized the publications department. He is now the managing editor of a computer magazine. In addition to managing, he coordinates a team of editors to produce a monthly publication.

Another graduate is with the nation's largest manufacturer of telephone equipment. Before she received her degree in technical writing, she had been a full-time homemaker with a Bachelor of Arts degree. She started out in the company working on internal news releases and helping engineers write research papers. She is now a public relations specialist and frequently presents papers at national conferences and contributes articles to professional journals.

Still another graduate has become one of the best-known teachers of technical writing in the country. After graduating with a Master's degree, he obtained his Ph.D. in communications. His interests range from communication theory to the application of electronics devices and from audiovisual aids to the teaching of technical writing.

Science institutes and research organizations always employ a number of technical writers. One woman is with the world's leading independent nonprofit scientific research and development institute as a senior writer in report and library services. Another is with the Los Alamos National Scientific Laboratory. A young man has become head of the Science Writers' Group with General Motors Research Laboratories.

## SUMMER EMPLOYMENT

It should be stated at the outset of this discussion that opportunities for students to obtain summer employment in technical writing vary according to the national economy. But there is a good chance that the well-trained student will be able to get a job. We have already described how some companies offer several kinds of summer jobs, including internships.

A technical communication student who worked during the summer for IBM described his work as follows:

> Our first big project was to edit and revise a manual describing the testing process used to warm up a computer for full-on operation. . . . As it was essential in this writing project to become thoroughly familiar with the equipment, we visited the site where the equipment had been designed and built and was being tested. We talked with the engineers on the project in order to completely understand the equipment. Finally, we took the existing manual, deleted much of the material from it, and added quite a lot of new material.

Students interested in this kind of short-term work must ask themselves a number of questions. Through my educational background, am I equipped to handle the products of science and technology? Am I a good enough writer to handle the communication phases? Am I located in, or is it possible for me to relocate to an area that needs technical writers?

Some people argue that summer employment in technical writing is a waste of time, that in such a specialized job, a major part of the summer may go by before the student really becomes productive. Nevertheless, if you can find a summer job in technical writing, by all means take it. It offers you the chance to get your foot in the door, giving you a decided advantage over other applicants with no experience who must start from scratch after they have graduated.

A summer job with Eastman Kodak, American Cyanamid, or any other company also gives you a chance to see if you are really suited for a technical writing career and whether the working conditions are what you expected them to be. Most employers will give you on-the-job evaluations of the work you have been doing and will let you know whether or not there may be a permanent job available after you graduate.

No organization, however, is willing to give you a free ride for the summer. You must demonstrate ability to get the work done, enthusiasm, and a desire to learn. Among the principal academic requirements today are a sound science background, competence in writing, and computer skills.

## RECENT DEVELOPMENTS

At a recent technical writing conference, one group meeting was opened with the following statement:

> The role of the technical communicator is changing. It behooves all old-time writers, editors, and illustrators to accept this and to prepare themselves for what will be expected of them.

New job categories in technical writing are constantly being developed. Many of these are highly specialized, requiring interesting combinations of skill, training, and aptitude. We will be exploring some new areas of endeavor in the following pages.

### Education

Colleges and universities are increasingly offering courses and programs in technical communication. Thus you should have no trouble finding the curriculum you want in a school near you.

Several trends in professional education can be seen. Companies and other organizations are showing greater interest in graduates with Master's degrees. An advanced program of this sort gives you the time to acquire more skills. It is frequently a step toward a supervisory or management job. Personnel interviewers tell us that they are always looking to the future: Do the candidates for a job have the characteristics and skills that will favor promotion into managerial positions? Companies are also looking for applicants who can handle worldwide documentation systems. It is now possible to sit at a console and send information all over the world. The technical writing student must be trained in the operation of these communication systems, on procedures in sorting out data and dealing with foreign countries that need the data.

If you examine the course listings of many colleges, you will find that an effort is being made to deal with what is called global information. Communication systems are now standard in many colleges, and courses in technical German, French, and other foreign languages are available. Most universities now offer computer competence training courses which take the "mystery" out of the computer. These courses have been designed for technical writers and others who will need to be computer literate.

## Word Processing

The word processor has made the job of the technical writer and editor more effective and productive. Most technical documents will be produced on the word processor, which has replaced the typewriter. Often these word processors are part of computer linkups with graphics and production, so that the technical editors can submit changes directly to the production department without ever leaving their desks. Less frequently will the editor be thought of as a lone individual blue-pencilling someone's manuscript. Now the text will be entered and stored in the word processor, and editing will be done using video display terminals. Since the word processor can store information, revising and updating documents is facilitated. Formats and paragraphs from earlier documents can be copied directly into the new document, thus eliminating much tedious typing, and additions can be inserted in the proper places without retyping pages. Word-processing programs and systems can index texts, incorporate proper headings, integrate visuals, check for spelling errors, update product names and numbers as well as client names, and turn out a clean copy as soon as the final editing task is completed. Combined with a laser printer, camera-ready copy often makes outside typesetting unnecessary. Proofreading and making copy changes or corrections can be done more quickly and efficiently, since no time is lost waiting for proofs to come from the typesetter. The word processor has indeed changed the technical writer's job and is certain to change it even more in the future.

## Machine Translation

Although this field hasn't experienced a complete breakthrough as yet, computer specialists are hard at work on the problem. If you have knowledge of one or more foreign languages and are skilled in technical writing, you should be well equipped to work with translating equipment to convert foreign languages into English and vice versa.

To give you some idea of how translators and technical writers may be assigned to machine translation, we contacted William L. Benzon of Rensselaer Polytechnic Institute. Dr. Benzon was formerly a bibliographer and editorial assistant with the *American Journal of Computational Linguistics*. This is what he said:

> As high technology spreads across the globe, the need for rapid, reliable, and relatively cheap translation of technical documentation grows proportionately. Translation, however, is slow, boring, but highly skilled work—which adds up to its being very expensive work as well. On the other hand, computers are fast and they don't get bored. If

they can be programmed with skills sufficient to the task, then fast and cheap translation may be possible.

Whether translation can be done by machines (MT) depends on the fact that most of the decisions to be made in translation are, in principle, as routine as the multiplication tables. Those decisions can be made quickly and accurately by a computer with the requisite software. However, many of those decisions depend on prior decisions of a different class, decisions which cannot be specified by some routine procedure. These decisions concern the meaning of the text and seem to require encyclopedic knowledge of the text's content. Current software provides literal, not idiomatic (everyday) language translations, so its use is restricted. Translated text must be converted into good idiomatic language, and there is an ongoing effort to solve this problem.

To illustrate the difficulty of idiomatic versus literal translation, consider the phrase: "the coast is clear." In Spanish, the equivalent phrase is "no haber ningunos moros en la costa" whose literal translation is "there are no Moors on the shore"!

Full-scale MT, in which the computer takes a text in the source language, such as Russian, and translates it into the target language, such as English, without any human intervention, may soon be possible. Software now exists which is capable of translating very short (300–500 words) texts on simple subjects, but these are too restricted in scope to be the basis of practical translation systems.

At present there are more practical machine aids to human translators. They have been given the designation MAT (machine-aided translation). Several MAT systems are in use by government agencies and private corporations. This subfield of technical communications will surely grow rapidly. In fact the technology in this area is changing so rapidly that what we have written may be obsolete by the time you read it. Machine translation is an exciting employment possibility for translators—

those people gifted in more than one language and with the ability to write well.

An acronym that we see today is CAT, computer-aided translation. In order to aid in the usage of CAT, companies are approaching translation in stages. First, a glossary, or vocabulary, of the most common technical terms used by that company for foreign translation is drawn up for the translators as well as for other writers. This glossary is entered in the computer memory. Then, a page of a manual, let us say, is entered in the word processor. It is pre-edited and then it is applied to the glossary. Words in the original piece of work are now transformed by the glossary, or "glossarized." After this is done, the piece of writing returns to the translator for its completion and final editing. Of course, this is an oversimplification of the process, but it may provide you with some idea of what is meant by CAT and machine translation. This development is by no means foolproof, but it does show the direction translation efforts are taking.

Systems of oral dictation to the computer are also being developed. Eventually, words spoken in one language may be computer-processed and translated into another language.

### Graphics and Multimedia

As companies compete to sell their products, they try out new techniques. Some of these are improved graphic presentations, the use of slides, movies, and other media devices. These media formats do not necessarily take the place of the written word; they complement each other. The result is that the technical writer is frequently called on to extend his technical communication skills to media communication.

In order for you to become as versatile as possible, our advice is to learn all you can about graphics and multimedia presentation. Take courses while you are in school; if you are already employed, contact the media people in your organization. Multimedia presentation is a fairly complex technique. In a small company you may be the only technical communicator. But don't think that your job will be confined to editing reports. You may be called on to prepare presentations for sales meetings, for speeches being delivered at professional meetings, and for in-company training programs. You should be prepared well beyond the necessary courses in writing.

### Film and Video Scripting

We are all familiar with these new methods of carrying infor-mation—audio and videotapes, cassettes, and films. Yet few of us are aware of the preparation that must be made before such media can be seen or heard. Someone has to prepare a script or scenario before it can be put into an electronic device, and many companies use technical writers for this job. And, as more devices for audio and video instruction are perfected, the involvement of the technical writer will increase.

An article in *Byline,* the Chicago Chapter newsletter of the Society for Technical Communication, entitled "Making Visuals Speak to Your Audience," focuses on the contribution of the script writer. Videos are an increasingly popular educational and train-ing tool. Mary Morse writes: "The technical communicator can quickly assume the role of the scriptwriter by using many of the skills used to develop tutorials, operating procedures, and user manuals. . . . The same steps used to develop technical documen-tation are also used to develop a video."

Just as a knowledge of photography is an asset in preparing articles for magazines, familiarity with film direction and produc-tion is important in the scripting of a video.

The need for technical writers should remain steady because technical writing is not a routine job. As we have pointed out, new developments in communication are continually taking place, and practically every industry now realizes that communication is the pipeline of American business. Whereas college placement offices in the past had little call for technical writers, requests are now appearing with increasing regularity. It is the rare personnel manager who is not on the alert for technical writers.

## The Computer and Documentation

The increased prominence of the computer has presented a new challenge for the technical writer—producing clear and usable computer-related documents. Lack of adequate documentation—that is, the written form of all the available information about a particular computer, computer program, or set of programs—is a major problem in modern industry. Documentation in its many forms—operating instructions, troubleshooting and repairs, user guides, etc.—is essential for management information on systems development and for proper coordination of subsequent phases of systems development and use. This documentation is often not thorough, nor is it done at the same time as the system is developed. Sometimes it is never done at all. The modern technical writer must be able to step into the complex documentation process and quickly and accurately be able to prepare such forms as the job run manual, the job control language manual, the balancing and control manual, the keyprocessing manual, and the job scheduling manual.

Obviously, special training is necessary for the technical writer to function effectively in preparing systems documentation. At a recent International Technical Communications Conference in Boston, Massachusetts, an internship training program offered by

Northeastern University which trains technical writers for the computer industry was described. The program offers computer science courses and training in writing operating instructions and programming reference manuals. This is just the beginning of the experience a technical writer must accumulate to function effectively in the area of systems documentation. At the same conference, a Honeywell Information Systems representative predicted that the technical writer will change to reflect the nature of new forms of documentation. He predicted that a whole new type of technical writer who is deeply involved in computer programming, and able to analyze data bases, and who is moreover familiar with software psychology, human factors, and ergometrics (the study of the ability of humans to adjust to their environment), will evolve to fill the communications needs of the computer age.

The preparation of manuals for people unfamiliar with data processing and programming is a continuation of what has long been a major role of the technical writer—bridging the information gap between the technical and the nontechnical person. Robert A. Ward of International Business Machines Corporation, in another presentation at the Boston ITCC, said that technical writers should move into program design, because they are best qualified to design the information package for the beginning user. IBM now hires technical writers to produce user manuals for their personal computers. The writer must learn what kinds of documents best fit the needs of the home computer buyer and what form would be best for these documents. This may require analyzing existing documents, interviewing users and designers, and generally converting technical and highly specialized language to language that the nonspecialist can understand.

# CHAPTER 6

# HOW TO GET STARTED

## EMPLOYMENT PROSPECTS AND JOB HUNTING

Employment for all kinds of writers and editors is expected to increase faster than other occupations through the year 2005. The use of salaried writers and editors by newspapers, periodicals, book publishers, and nonprofit organizations will grow along with growing demands for the publications. The demand for technical writers should even exceed the demand for writers in general because the amount of technical and scientific writing continues to increase at an explosive rate. The departure of people for other employment or retirement will create additional job opportunities.

Through the year 2005, the outlook for most writing and editing positions will continue to be very competitive because many people are attracted to this field. Opportunities for technical writers will remain good because only a small number of people can interpret technical material and make it understandable to the general public.

In the last few years, a considerable amount of information has been offered about how to get started in technical writing, in the form of ads in the newspapers and journals, brochures prepared by the professional societies, and even a few books. If you are seriously thinking about becoming a technical writer, you can take a number of steps that will help you obtain professional guidance and information. Get in touch with the Education Committee of the Society for Technical Communication. The Committee exists to inform people on how to prepare themselves for the profession. It is ready to answer your questions and will send you the names of prominent members of the Society to whom you may write for advice.

The Society has various publications available on various aspects of technical writing: see Appendix A. Every year, at the close of the annual international conferences, STC publishes *Proceedings,* which includes the complete text of all the papers presented at the meeting. This publication can be obtained at STC headquarters and from many technical writing teachers. The *Proceedings* is a valuable tool in preparing yourself for a career because it contains hundreds of pertinent papers delivered by professionals.

If you are still in high school, make an appointment with your guidance counselor to discuss the profession of technical writing. A lot depends on whether you are planning to continue your education by going to college or planning on other specialized training courses. In either case, counselors should have literature available about technical writing careers or tell you where it can be obtained.

If you are in college, talk with the official in charge of the placement office. Job placement is a service provided by almost every institution these days. Throughout the year, college placement officers are in contact with the personnel managers of companies and other organizations that are looking for people to fill important technical writing jobs.

But don't depend entirely on the college placement office. Throughout the academic world of technical communications are many exceptionally competent teachers, too many to list here. If there is a technical writing program in your college, there will most likely be a knowledgeable person teaching it. And nine times out of ten, this person will have good contacts with business and industrial firms. Contacts of this kind are probably the most valuable way of getting started in the profession.

Job contacts can also be established by getting in direct touch with the supervisors and administrators of the publications departments of the companies themselves. To establish these contacts, read the large industrial ads for technical writers in the newspapers, especially those in highly developed industrial areas. If you can't find a specific name to send your inquiry to, send it to the director of publications. In time, your letter will filter through to the right person.

## RESEARCH YOUR JOB

If you wish to enter the industrial world, or any other occupation for that matter, you should do some groundwork. You should first assess what you have to offer. Then do some research into the company in which you are interested to see whether you can meet their requirements.

To some extent you can do your own research. And remember that certain kinds of companies—electronic, aeronautical, and chemical industries, contracting companies, government agencies, and research institutes—are more in need of technical writers than others.

Most libraries in your community will have a copy of *Standard and Poor's Index* on their shelves. This reference book lists a great deal of information about major companies—where they are located, what they manufacture, divisions in the company, and branches in various cities and countries.

There are other helpful books and pamphlets you can consult which give valuable career information. Plan to spend some time in your local public library, university library, or a community college library getting the information you need.

During the last decade, employment agencies dealing exclusively in the industrial employment of engineers and administrators have grown. These agencies are now adding technical writers to their lists.

In this respect, we did some research of our own on your behalf. Here are some of our findings taken from ads in various journals:

- Editor for consumer electronics. Must be able to use production techniques.
- Senior editor for trade magazine. $30,000–$38,000.
- Editor for medical journals; monographs on clinical medicine.
- Editor, nursing journal, $28,000.
- Editor/writer to report on technology in robotics field. Needs technical background.
- Editing/writing skills for medical journal. Will teach use of word processor—$18,000.
- Business writer for management consulting firm. Must be comfortable with semitechnical subjects. High, $36,000.
- Editor to coordinate production of proposals for computer services. Experience with government proposals.
- Newsletter editor for part-time work in university for space research group.

*Technical Communication,* the STC journal, is one of the best sources for employment opportunities. Ads appear in every issue. Get into the habit of reading the magazines and newspapers that feature industrial news to learn the names of such agencies.

## LETTER OF APPLICATION

Once you have the name of a potential employer, you are in a position to write a letter of application with which you will include a résumé of your accomplishments if specifically requested. There are various ways to write a letter of application, but it is preferable not to include a résumé or to mention salary expectations in your introductory letter.

The application letter should be short, but it should include information about your background and the type of position you are seeking. If you have a mutual friend or professional contact who has suggested you write to the company, you should mention it. A reference to an ad in a newspaper or magazine frequently will help to establish the right atmosphere. If you exhibit in your letter a tone of enthusiasm, a knowledge of the requirements of the company to which you are applying, and adequate training, your letter will most certainly be read.

## RÉSUMÉ AND INTERVIEW

Here are some comments before we give you specifics about the résumé. It should be a professional piece of work—remember, you are applying for a job as a writer. Therefore, every sentence you write should demonstrate your ability. This is no place for sloppy typing, poorly constructed paragraphs and sentences, or misspellings.

Because the responsibilities of employers toward their employees are changing drastically, here are several things that you may wish to omit from your résumé, or at least be cautious about. Religious affiliations: only in rare cases would this improve your chance of getting a job; race: antidiscrimination laws prevent questions in this category; sex: employers usually can tell the sex of an applicant by the name; certain personal data such as age, marital status, children, which have no bearing on your ability to do the job.

Both the résumé and the letter of application should be of the highest professional caliber, and the following information should be included.

*Type of Work Sought.* This section lets the prospective employer know that you have given some thought to your future, that you have assessed your work skills, and that, above all, you understand what technical writing and editing involves.

*Previous Employment.* The prospective employer is interested in seeing the complete employment picture. Any summer or part-time employment is indicative of a willingness and an ability to work.

*Education.* Start with your most recent education and work backward. It is not necessary to go further back than your high school education. However, be sure that you include here any particular programs with which you have been associated. A course in statistics taken at a summer school may be the one thing that sells you to the employer as you compete with other applicants.

*Courses.* Divide these into graduate, undergraduate, and high school courses. Put the latest group first to bring the reader up to date quickly. List only those courses that you feel are pertinent to the job you are seeking. These may include writing, oral presentations, science, engineering, computer training, and languages. Specific courses are less important if you have relevant practical experience or advanced degrees.

*Honors and Offices.* Every company likes to think that it can spot a winner. One indication of this is the activities you participated in when you were in school, especially those relating to your chosen profession. All academic honors and scholarships should be cited. If you are a recent graduate and have a B+ or higher grade point average, it should be included.

*Professional Organizations.* This will apply more to those who are changing professions or seeking new jobs.

*References.* References are important because they will show the employer how professionals evaluate your skills and the kind of work experience you have had. References, therefore, should include three or four teachers and employers in your professional field. References from family friends who have no gauge of your professional competence are not appropriate. Always contact your references first, asking permission to use their names.

Almost any current textbook on technical writing can provide you with a sample letter of application.

## SALARIES

The salaries offered to students seeking their first jobs in technical writing cannot be definitely established; however, certain basic principles do apply:

- With a Bachelor's or Master's degree in engineering or science, a student can command a higher beginning salary than a student with a degree in English or some other nontechnical subject.
- Students graduating from certain prestigious colleges usually can command higher salaries than students from lesser known schools.
- The higher the course grades, the more summer experience obtained, and the more the student can display characteristics of ability and initiative, the higher the salary is likely to be.
- Technical writers with degrees in certain areas, notably electrical engineering and electronics, are in greater demand than students with training in other areas.

Beginning technical writers are likely to be evaluated very closely on the basis of their educational records, their writing ability, and their potential for being promoted. A survey made within several technical colleges, relating to recent graduates pursuing technical writing programs, revealed these average starting salaries:

| Bachelor's degree *Nontechnical* | Bachelor's degree *Technical* | Master's degree *Both* |
|---|---|---|
| $18,000–20,000 | $21,000–23,000 | $25,000–30,000 |

You can assume that if you graduate from a college or university with some expertise in technical writing, you will be within a few hundred dollars of these median salaries.

It is apparent that the highest salaries are offered in industry—especially in the electronics and computer fields. Lower salaries can be found in the technical press and in other categories where an engineering background is not necessary. Starting salaries can be expected to increase about 3% annually.

## SAMPLE RÉSUMÉ

George B. Anthony
4 State Street
Mason, New York 12189

POSITION SOUGHT:
Technical writer/editor in publications or public relations.

EXPERIENCE:

| *Organization* | *Position* | *Dates* |
|---|---|---|
| *Journal of Petrology*, Bath College of Science | Assistant to the Editor | July 1992– present |
| National Retail Stores | Worked in the shipping department | July & August 1985/1986 |

EDUCATION:
Bath College of Science
M.S. in Technical Writing
Important Courses: Writing for Publication, Writing and
Editing, Technical Writing in Industry, Graphics, Public
Relations, Data Processing, Introduction to Computers,
Scientific German
Pennsylvania Institute of Technology
B.S. in Geological Sciences
Important Courses: geology, physics, mathematics, chemistry,
technical writing, effective speech, German, anthropology.

COLLEGE ACTIVITIES:
Secretary, student chapter, National Petrology Society
Features Editor, *P.I.T. Engineering Magazine*

HONORS:
Tuition scholarships, junior and senior years, Pennsylvania
Institute of Technology
Fellowship at present, Bath College of Science
References available upon request.

# CHAPTER 7

# PROFESSIONAL ASSOCIATIONS

The organizational history of the technical communication profession is an interesting one. Some years ago, small groups of people in different parts of the country gathered together to chart a course for a fledgling career that seemed to be on its way to becoming an important profession.

By March 1955, technical writers from the Boston area had organized themselves to such an extent that they felt the time had come to draw up a constitution. This was the Society of Technical Writers, which by that time had informal chapters in Connecticut, Illinois, Maryland, Massachusetts, New Jersey, New York, Pennsylvania, and Tennessee.

From that time on, the Society of Technical Writers flourished. In June 1954, it had put out its first publication, called the *Technical Writing Review*. The concerns of that early society are mirrored in a statement of aims published in the first issue:

1. Developing and establishing standards for technical writing.
2. Stimulating the exchange of information of common interest in this and allied fields.
3. Encouraging the development and training of technical writers.
4. Acquainting others with the profession.

About the same time, a group of technical writers was getting together in the New York City area. The early days of this group were described in a bulletin published by this new organization, the Association of Technical Writers and Editors:

> Although those in the various technological communica-
> tion fields have long felt the need for organization, it
> remained for the 1953–54 period to bring the movement
> into noticeable activity throughout the country. This activ-
> ity culminated in the formation of at least five different
> writing societies. The centers of the organizations are Bos-
> ton; Washington; Los Angeles; Oak Ridge, Tennessee; New
> York; and London, England—as far as is known. The
> Association of Technical Writers and Editors was the New
> York group. It was organized by individuals interested in
> the improvement of technological communication on all
> levels, and it is the only technical writing society in New
> York.

The objectives of this group as stated in the bylaws were to:

1. Define, establish, and observe standards of technological writing, editing, and publishing.
2. Uphold and justify the self-respect and professional stand-ing of persons engaged in the several domains of techno-logical communication.
3. Facilitate intergroup and interindividual contact and discus-sion among those interested in problems of technological communication.
4. Encourage the development and training of technical writ-ers.
5. Acquaint others with the profession of technical writing.

On the West Coast, the same professional stirrings were taking place, due in part to the thriving chain of industry along the West Coast, consisting of aircraft companies, electronics companies, research institutes run by and for the government, military instal-lations, and academic research institutes. The geographical location of these organizations, separated from the East Coast, was largely responsible for the founding of the Technical Publishing Society. Its aims were similar to those of the two East Coast groups.

In 1957 the two East Coast societies merged to form the Society of Technical Writers and Editors (STWE) and their respective publications were also combined to form the *STWE Review*. The new professionalism of this group was emphasized by a statement in the *Review*:

> The primary objective of the *STWE Review* will be to contribute to the professional advancement of technical writing and editing.

## SOCIETY FOR TECHNICAL COMMUNICATION

Today the official society of the technical writing profession is the Society for Technical Communication (STC). In a prospectus describing its activities, the Society makes this statement:

> The Society for Technical Communication is the world's largest professional organization dedicated to the advancement of the theory and practice of technical communication in all media. Members in more than fifty chapters and branches typically are engaged as technical writers and editors, graphic artists, educators, publications and communications managers and directors, technical presentation specialists, engineers, designers, technical communication consultants, marketing specialists, technical and scientific publishers, audiovisual specialists, information science researchers, documentation specialists, and advertising copywriters.

> Through chapter meetings, ideas are exchanged, programs benefiting the individual members are presented, and the members are given the opportunity to express their

views. Some chapters have been very active in their industrial communities in bringing business and education together, presenting important speakers and developing writing workshops.

The Society's membership is kept in touch through its journal, *Technical Communication*. It contains articles dealing with the latest developments in all phases of technical communication, together with reviews of new books and news of noteworthy events.

The other STC publication is a newsletter, *INTERCOM,* which acquaints chapters and individual members with internal activities: records of meetings, items concerning individuals and chapters, formulation and news of committees, and changes in rules and regulations. Many of the local chapters also publish newsletters.

The society offers a low cost membership for students and they publish a quarterly newsletter for their student members.

The STC publishes books and pamphlets which are authoritative and arise from practical experience. They are often collections of contributed articles from their journals (see Appendix A).

The STC offers a variety of publications to help novice and experienced technical communicators develop and enhance their skills. Some of the titles in their series are: *Basic Technical Writing, Academic Programs in Technical Communication, Technical and Business Communication, The Scientific Report: A Guide for Authors, Freelance Nonfiction Articles,* and *Guide for Preparing Software User Documentation.*

Whether you are a student or a practicing technical writer, you should plan to attend the annual conferences of STC, held each spring in various locations. A recent conference attracted over 2000 members and speakers from this country and several European countries. Awards were given, members honored by being advanced to Fellows, and a multitude of subjects—writing, editing, graphics, organization—presented. At these conferences, you will learn about communication firsthand—from professionals, administrators, and teachers. Discussions with professionals working in the field will allow you to better understand what technical writing is and in what kind of atmosphere the technical writer operates.

To become a member of the Society for Technical Communication, you must be employed or be concerned with the profession of technical writing and publishing. The qualifications and requirements are best summed up in the Society's brochure, which can be obtained from:

Society for Technical Communication (STC)
901 Stuart Street, Suite 304
Arlington, VA 22203

There are many STC chapters throughout the United States, and you can get information about these from the above address.

## RELATED FIELDS

Technical writing is only one part of a larger picture that you might call professional writing—communication skills applied to such professions as medicine, pharmaceuticals, public health, and business. The devices used to transmit information in any of these specialties are similar to those used in other fields, but of course the information is different, as is the vocabulary.

Two kinds of professional communication bear a great resemblance to technical writing: medical and business writing. Each has its own professional society, the American Medical Writers Association and the Associated Business Writers of America (addresses in Appendix B). Some members in each of these organizations belong to both, as well as to the Society for Technical Communication. Because of their common characteristics, what we have already said about technical writing can certainly be applied to medical and business writing.

## Medical Writing

Here is the professional history of one medical writer. She graduated from a college in Virginia with a Bachelor's degree in psychology and English and later earned a Master's degree in psychology. One of the large Army medical centers had grown to a point where it needed a technical publications editor. The center was issuing an increasing number of scientific papers and was about to publish its own journal. The young woman was hired to fill this new job. Out of this grew a more challenging position in an Army research institute. In addition, this medical writer has been made an associate editor of a journal of sports medicine and an editorial consultant to the Eisenhower Medical Center.

Another writer has a Bachelor of Arts degree in journalism. For a time he was editor of a trade publication and later became an editor for a publishing firm producing technical books and pamphlets. He then joined a state health department and became an editor on the department's monthly magazine. Among his present duties are writing and editing pamphlets, booklets, and brochures for all units within the department; preparing the department's annual report; and preparing news releases concerning department activities and providing information for the press. This man classifies himself as a medical writer, or in a narrower sense, a public health writer.

Many physicians have become medical writers, just as many engineers have become technical writers. Here is a case in point: A writer received a degree in pharmacology from one college, a Bachelor of Science degree from another, and a Medical Degree from a third. He has been both a practicing pharmacist and a physician. He has served as medical director and director of research for a large chemical manufacturing company and is now with a New York advertising firm as medical director in charge of clinical research. In other words, it is the job of his department to delve into the research and manufacture of the drugs and chemicals of his client firms so that they may be advertised intelligently and accurately in the medical journals.

Or you might be inclined to a career in medical journalism. If you go into a medical library, you will be overwhelmed by the number of journals displayed there. There are literally thousands of them—in medicine, dentistry, veterinary medicine, public health, and research and development. What we have already said about technical journalism in previous chapters pretty well applies to medical journalism. If you join the staff of the *Journal of the American Medical Association* or a more general magazine like *Today's Health,* you will be writing articles, editing other people's work, and acting as a liaison with artists and production personnel.

In medical writing there are many free-lance writers. Many of these writers work at home on manuscripts and reports originally written by doctors and researchers who are too busy to put on the finishing touches. These free-lance writers are frequently highly qualified scientists who wish to do only part-time work. So, if you are trained in biology, chemistry, or in one of the more specialized sciences, you may wish to consider this very important kind of medical writing. It may be just the right kind of employment for you.

According to Dr. Eric W. Martin, a physician and communicator, the duties of medical editors and writers may include the following:

- *Clinical Brochures.* These are publications containing the data gathered together in a company from years of research on new drugs.
- *Case Report Forms.* With these forms, medical writers help clinical investigators report on the medical histories of patients.
- *Clinical Research Reports.* Medical writers prepare these summaries of research and clinical data which are important

in determining whether drugs will be available for patient use.

- *Clinical Papers.* In the preparation of these papers, the medical writers give editorial and writing help to busy scientific investigators.
- *Physician Brochures.* Doctors must know about the drugs they prescribe for their patients. These brochures specify dosages, side effects, and other pertinent information.
- *Official Brochures.* These are commonly called package circulars or package inserts, again for the benefit of the physicians.
- *Abstracts.* Medical writers screen hundreds of important journal articles and condense them for scientists and researchers.
- *Guides.* The manufacturers of medical products must follow strict federal and state government regulations. Medical writers frequently prepare manuals to interpret these controls for company employees.
- *Media Preparation.* Medical writers coordinate the preparation of motion picture scripts, slide presentations, and cassette courses.

If you have a strong background in chemistry or biology, investigate the possibility of working in the pharmaceutical rather than the chemical industry. If the ultimate consumer of the company's products is the public, then as a communicator for such a firm you would be classified as a medical writer.

Or you may find a niche for yourself in medical advertising. Just as there are many technical advertising firms and public relations firms, so there are many advertising companies that specialize, at least in part, in drawing the attention of the public and of physicians to the availability of new drugs.

The aims of medical writing can be summed up by a brief description of the American Medical Writers' Association

(AMWA), which was organized under its present name in 1948. The Board of Trustees has issued this statement on the subject:

> The aim of the AMWA is to bring together into one association all North Americans who are concerned with the communications of medicine and allied sciences in order to maintain and advance high professional standards. ... Its purpose is educational, scientific, and literary. By providing an annual meeting addressed by distinguished authors, editors, and teachers, an opportunity is afforded to all interested in any phase of medical communication to keep themselves informed of the progress being made to maintain and advance high professional standards and thus to aid in general medical advancement.

Of interest to students will be the collegiate training sponsored by the AMWA in medical journalism and the AMWA scholarships. According to the latest information, the Schools of Journalism at the University of Illinois at Urbana, the University of Missouri at Columbia, and the University of Oklahoma at Norman have established courses in medical journalism. The AMWA itself has established a Medical Journalism Scholarship Fund. Through this fund, the AMWA offers scholarship assistance to needy and talented students.

The journal of AMWA is the *American Medical Writers' Association Journal,* published four times a year. Here are a few representative articles in recent issues: "Hippocrates—First Medical Writer," "Writing Persuasive Proposals," and "Dissemination of Scientific Information from the Author to the Public."

Membership in AMWA is open to persons actively engaged or interested in any aspect of communication in the medical and allied professions and services. For further information, write:

American Medical Writers' Association (AMWA)
9650 Rockville Pike
Bethesda, Maryland 20814

## Business Writing

If you elect a career in business writing, you may be dealing with readers ranging from business managers to the general public, and with such topics as personnel and consumer relations. Several associations for business writers are listed in Appendix B. The director of the Association for Business Communication (ABC) has stated that a person wishing a career in business and industry should study business and the communication of business. The kind of writing or editing one trains for should be selected carefully, as business writing is highly specialized. It is one thing to write repair manuals and a quite different thing to write training materials for the sales personnel. A portfolio of samples should be accumulated by aspiring business writers and they should learn to write good letters of application to accompany their professional portfolio.

Business writers can fall into a number of classifications. The following list is from an article in the *ABC Bulletin:*

Informal auditor—performs operational reviews and reports on recommendation for management.

Financial analyst—does analytical reporting.

Accountant—writes company policies and procedures.

Researcher (advertising)—organizes and writes final reports; makes formal presentation to clients.

Product advertiser—plans and coordinates product development; composes literature and sales aids.

Community planning specialist—writes, edits, and reports.

Planning director—oversees and prepares copy for promotional literature.

Publications specialist—writes original copy, graphic arts, and layouts.

Corporate relations officer—writes news releases and edits materials.

Proposal specialist—plans, writes, and produces contract proposals.

This is only part of the list, which includes job descriptions for technical and medical writers. In all the descriptions, you will find that, once again, there is an overlapping of job duties among business, medical, and technical communicators.

A report prepared by a committee of ABC discussed business communication education in the United States. The committee had this to say about undergraduate education:

> At Baylor University's Hankamer School of Business, Department of Business Communications and Business Education, students may receive a BBA degree in business communications. That major is narrowly defined and directed at persons who are preparing to be executive and administrative secretaries. Central to the degree are two core courses: integrated business writing and speech for professional people. Other required courses are word processing; shorthand, dictation, and transcription; and advanced office procedures. Thus the field is for persons planning a career in the secretarial field.
>
> A more traditional BBA degree with a major in business communication is offered at Western Michigan University, College of Business, Department of Business Education and Administrative Services. There the BBA in business requires twenty-four hours for a major and eighteen hours for a minor. Their program includes the following topics:
>
> Information Writing
> Business Communication
> Organizational Communication
> Teaching Internship
> Internship (individual, nonteaching)
> Advanced Business Writing
> Report Writing
> Topics in Business Communication (topic changes)
> Independent Study
>
> For thirteen years, the University of Southwestern Louisiana, College of Commerce, has offered a Bachelor of

Science degree in business administration with a major in business communication. The department offers four undergraduate courses: business letter writing, communication in business, business research and reporting, and independent study. Remaining requirements for the degree come from areas such as behavioral management, journalism, speech, English, and psychology.

The ABC committee also discussed graduate education. It stated that several schools offer the traditional Master of Business Administration core courses with students permitted to concentrate on communication. The University of Texas at Austin includes core business courses, as well as such optional communication courses as the job-getting process, business report writing, behavioral communication, and communication research for organizations.

The following pattern is typical: MBA students meet core requirements in accounting, marketing, statistics, operations research, and finance. They may also select business communication courses as electives, either within the business school if such courses are offered, or in another school of the university. Such latitude was evident at both the graduate and undergraduate levels at many schools.

ABC is the professional organization for business writers. It has well over a thousand members drawn from teachers of business communication and business writers.

The aims of ABC are similar to those expressed by STC and AMWA—to provide a clearinghouse for its members; in addition, it promotes professional standards on the job. ABC carries out its purpose through an annual national convention. Throughout the year, regional meetings and seminars are offered.

Two journals are published by ABC: the *Journal of Business Communication,* which features articles on research results, approaches to business communication, and techniques of wide application; and the *ABC Bulletin*, which contains a wealth of practical material. ABC also publishes a number of reference books, including *Technical Writing: A Bibliography.*

If you should be thinking of business writing as a career, you will be interested to know that student chapters of ABC have been developed at ten colleges under the general title of Alpha Beta Chi. You will find chapters at Western Michigan University, Creighton, Bentley, Florida Institute of Technology, Bowling Green, Oklahoma State, Macomb County Community College, University of Minnesota, St. Paul's, and the University of Washington. For further information about business writing careers, write:

Executive Director
ABC
Department of Management
College of Business
University of North Texas
Denton, Texas 76203

## OTHER PROFESSIONAL WRITING GROUPS

We have explained that technical writing and other forms of technical communication are at the center of an expanding field in which the principles of technical writing are carried over into other professions.

There are a number of professional societies devoted to more specialized forms of technical communication. Their names should be self-explanatory. Additional organizations are listed in Appendix B at the back of this book.

Aviation/Space Writers' Association
17 S. High Street
Columbus, OH 43215

Council of Biology Editors
111 E. Wacker Drive
Chicago, IL 60601

National Association of Science Writers
P.O. Box 294
Greenlawn, NY 11740

IEEE Professional Communication Group*
345 E. 47th Street
New York, NY 10017

National Association of Government Communicators
609 S. Washington Street
Alexandria, VA 22314

*Institute of Electrical and Electronics Engineers.

## TECHNICAL WRITING JOURNALS

The profession of technical writing has grown to the extent that most of the societies, foreign and domestic, publish journals and newsletters. In addition, two other journals have appeared during the past few years.

One of these is the *Journal of Technical Writing and Communication*. It expresses the views of communicators, records their achievements, helps to promote their research, accepts papers from students, and in general acts as a forum for professional activities. Information about this journal can be obtained from:

Baywood Publishing Company
20 Austin Avenue
P.O. Box 337
Amityville, NY

*The Technical Communication Quarterly,* formerly *The Technical Writing Teacher,* is concerned primarily with the teaching aspects of technical writing; it fills a need by circulating methods and instruction among people entrusted with developing curricula and programs. If at some time you are called upon to teach a course, whether in college or in industry, the articles in *The Technical Communication Quarterly* will be of great assistance to you. The journal's sponsoring group is The Association of Teachers of Technical Writing. Inquiries should be sent to:

University of Minnesota
Rhetoric Department
202 Haecker Hall
St. Paul, MN 55108

Illustrating the continuing growth of technical writing, both domestically and internationally, is the fact that the Society for Technical Communication (STC) has become an international organization with over a thousand members living and working outside of the United States. Their newsletter, *INTERCOM,* is mailed to many foreign addresses, since a number of their 150 chapters are located in other countries, including Canada, Great Britain, France, Italy, Israel, Japan, and Taiwan to name a few. Several international organizations with U.S. headquarters are listed in Appendix B, as are a number of associations based in Canada.

# BIBLIOGRAPHY OF
# RECOMMENDED REFERENCES

## Books

Alley, Michael. *The Craft of Scientific Writing*. New York: Prentice Hall, 1988.

Anderson, Paul V. *Technical Writing: A Reader-Centered Approach*, 2nd ed. Fort Worth, TX: HBJ College Pubs., 1990.

Beene, Lynn D., and Peter White. *Solving Problems in Technical Writing*. New York: Oxford University Press, 1988.

Beer, David F., ed. *Writing and Speaking in the Technology Professions: A Practical Guide*. Piscataway, NJ: IEEE, 1992.

Bell, Paula. *Hightech Writing: How to Write for the Electronics Industry*. New York: John Wiley, 1985.

Bernstein, Theodore M. *The Careful Writer's Guide to the Taboos, Bugbears, and Outmoded Rules of English Usage*. New York: Simon & Schuster, 1991.

Bittner, John R. *Mass Communication: An Introduction*. Englewood Cliffs, NJ: Prentice Hall, 1989.

Blicq, Ron S. *Writing Reports to Get Results: Guidelines for the Computer Age*. New York: IEEE Press, 1987.

Bowers, Bege K., and Chuck Nelson, eds. *Internships in Technical Communication*. Soc. Tech. Com., 1991.

Bowman, Joel P., and Bernardine P. Branchow. *How to Write Proposals that Produce*. Phoenix, AZ: Oryx Press, 1992.

Brusaw, C. T., G. J. Alfred, and W. E. Oliv. *Handbook of Technical Writing*. New York: St. Martin's Press, 1987.

Butcher, Judith. *Copy-Editing*, 2nd ed. Cambridge, England: Cambridge University Press, 1983.

Chandler, Harry E., ed. *Technical Writer's Handbook*. Materials Park, OH: ASM International, 1983.

*The Chicago Manual of Style,* 14th ed. Chicago, IL: University of Chicago Press, 1993.

Clements, Wallace, and Robert Berlo. *The Scientific Report: A Guide for Authors*. Arlington, VA: Soc. Tech. Com., 1984.

Deen, Robert. *Business Communications*. Lincolnwood, IL: NTC Publishing Group, 1987.

Emerson, Francis B. *Technical Writing*. Boston, MA: Houghton Mifflin, 1987.

Ensley, Therese, ed. *Marketing Yourself as an Independently Employed Professional*. Arlington, VA: Soc. Tech. Com., 1991.

Estrin, Hermann. *Technical Writing in the Corporate World*. Los Altos, CA: Crisp Pubns., 1990.

Farkas, David. *How to Teach Technical Editing*. Arlington, VA: Soc. Tech. Com., 1987.

Fear, D. E. *Technical Communications,* 2nd ed. Glenview, IL: Scott Foresman, 1981.

Foote-Smith, Elizabeth. *Writing*. (covers all types of writing careers). Lincolnwood, IL: NTC Publishing Group, 1984.

Gibson, Martin L. *Editing in the Electronic Era,* 3rd ed. Ames, IA: Iowa State Press, 1984. *The Writer's Friend: for Copy Editors and Others Who Work with Publications,* Ames, IA: Iowa State Press: 1989.

Goodman, Neville W., and Martin B. Edwards. *Medical Writing: A Prescription for Clarity*. New York: Cambridge University Press, 1991.

Harkins, C., and D. L. Plung, eds. *A Guide for Writing Better Technical Papers*. New York: IEEE Press, 1982.

Hart, Andrew W., and J. Reinking. *Writing for College and Career,* 4th ed. New York: St. Martin's Press, 1990.

Horton, William. *Secrets of User-Seductive Documents*. Arlington, VA: Soc. Tech. Com., 1991.

Houp, Kenneth W., and T. E. Pearsall. *Reporting Technical Information,* 6th ed. New York: Macmillan, 1988.

Huth, Edward J. *How to Write & Publish Papers in the Medical Sciences,* 2nd ed. Baltimore, MD: Williams & Wilkins, 1990.

Kent, Peter. *Technical Writer's Freelancing Guide*. New York: Sterling, 1992.

Laster, A. A., and N. A. Pickett. *Occupational English,* 4th ed. New York: Harper & Row, 1990.

Lawrence, N. R., and E. Tebeaux. *Writing Communications in Business and Industry,* 3rd ed. Englewood Cliffs, NJ: Prentice Hall, 1982.

Lay, Mary M., and William Karis, eds. *Collaborative Writing in Industry: Investigations in Theory & Practice.* Los Gatos, CA: Baywood Publishing, 1991.

Lefferts, Robert. *Getting a Grant in the 1990s: How to Write Successful Grant Proposals.* New York: Prentice-Hall, 1991.

Leonard, Donald J., and Shurter, Robert L. *Effective Letters in Business,* 3rd ed. New York: McGraw-Hill, 1984.

Lock, Stephen, ed. *A Difficult Balance: Editorial Peer Review in Medicine.* Philadelphia, PA: ISI Press, 1986.

Locke, David. *Science as Writing.* New Haven, CT: Yale University Press, 1992.

McDowell, Earl E., ed. *Interviewing Practices for Technical Writers.* Amityville, NY: Baywood Pub., 1991.

McQuaid, Robert W. *The Craft of Writing Technical Manuals,* 2nd ed., (with workbook). San Diego, CA: R. W. McQuaid, 1990.

Markel, Michael H. *Technical Writing: Situations & Strategies.* New York: St. Martin's Press, 1992.

Mathes, J. C., and D. W. Stevenson. *Designing Technical Reports,* 2nd ed. Indianapolis, IN: Bobbs-Merrill, 1991.

Michaelson, Herbert B. *How to Write and Publish Engineering Papers and Reports,* 2nd ed. Philadelphia, PA: ISI Press, 1986.

Miller, Diane F. *Guide for Preparing Software User Documentation.* Arlington, VA: Soc. Tech. Com., 1988.

Mills, Gordon H., and J. A. Walter. *Technical Writing,* 5th ed. Fort Worth, TX: HBJ College Pubs., 1986.

Murphy, Herta, and Herbert W. Hildebrandt. *Effective Business Communications,* 5th ed. New York: McGraw-Hill, 1988.

O'Connor, Maeve. *How to Copyedit Scientific Books and Journals.* Philadelphia, PA: ISI Press, 1986. *Writing Successfully in Science.* New York: Unwin Hyman (Chapman & Hall), 1992.

Pauley, Steven E., and Daniel Riordan. *Technical Report Writing Today,* 4th ed. Boston, MA: Houghton Mifflin, 1989.

Pearsall, Thomas E., and Donald H. Cunningham. *The Fundamentals of Good Writing.* New York: Macmillan, 1988. *How to Write for the World of Work,* 4th ed. Fort Worth, TX: HBJ College Pubs., 1990.

Pearsall, Thomas E., and Frances J. Sullivan. *Academic Programs in Technical Communication.* Arlington, VA: Soc. Tech. Com., 1985.

Perdue, Lewis. *The High Technology Editorial Guide & Stylebook with Disk,* (available for IBM or Mac). Homewood, IL: Busn. One Irwin, 1991.

Pickett, Nell A., and Ann A. Laster. *Technical English Writing, Reading and Speaking,* 5th ed. New York: Harper & Row, 1989.

Plotkin, Helen, and Carole Mablekos, eds. *Technical Writing and Communication,* (includes 6 cassettes). Washington, DC: Am. Chem. Soc., 1988.

Plotnik, Arthur, *The Elements of Editing: A Modern Guide for Editors and Journalists.* New York: Macmillan Publishing Co., 1986.

Rook, Fern. *Slaying the English Jargon.* Arlington, VA: Soc. Tech. Com., 1992.

Sachs, Harley L. *Freelance Nonfiction Articles.* Arlington, VA: Soc. Tech. Com., 1987.

Schoff, Gretchen H., and Patricia A. Robinson. *Writing & Designing Manuals: Operator and Service Manuals, Manuals for International Markets.* Boca Raton, FL: Lewis Pubns., 1991.

Schwager, Edith. *Medical English Usage & Abusage.* Phoenix, AZ: Oryx Pr., 1990.

Shaw, Harry. *Punctuate It Right.* New York: Harper & Row, 1993.

Sigmund, N. B. *Communication and Business,* 3rd ed. Glenview, IL: Scott Foresman, 1982.

Sims, Brenda R. *A Handbook of Examples.* Denton, TX: University of North Texas Press, 1990.

Smith, Peggy. *Mark My Words: Instruction and Practice in Proofreading.* Alexandria, VA: Editorial Experts, 1987.

Souther, James W., and L. White. *Technical Report Writing,* 2nd ed. Melbourne, FL: Krieger, 1984.

Slatkin, Elizabeth. *How to Write a Manual.* Berkeley, CA: Ten Speed Press, 1991.

Stewart, Ann. *Technical Writer.* Fort Worth, TX: HBJ College Pubs., 1988.

Stoughton, Mary. *Substance and Style: Instruction and Practice in Copyediting.* Alexandria, VA: Editorial Experts, 1989.

Tarutz, Judith. *Technical Editing: The Practical Guide for Editors and Writers.* Redding, MA: Addison Wesley, 1993.

Tibbetts, Ann. *Practical Business Writing.* Glenview, IL: Scott Foresman, 1987.

Treece, Malra. *Successful Communication for Business & the Professions,* 4th ed. Needham Heights, MA: Allyn & Bacon, 1990.

Treece, Malra, and Larry Harman. *Effective Reports for Managerial Communication,* 3rd ed. Needham Heights, MA: Allyn & Bacon, 1991.
Turk, C., and J. Kirkman. *Effective Writing: Improving Scientific, Technical & Business Communication,* 2nd ed. New York: Chapman & Hall, 1989.
Van Buren, R., and M. F. Buehler. *The Levels of Edit,* 2nd ed. Washington, DC: U.S. Government Printing Office, 1980.
Van Wicklen, Janet. *Technical Writing Game: A Comprehensive Career Guide for Aspiring Technical Writers.* New York: Facts on File, 1992.
Waters, Max L., and Don Leonard. *Teaching Business Communication.* Little Rock, AR: Delta Pi Epsilon, 1985.
Young, Matt. *Technical Writer's Handbook: Writing with Style and Clarity.* Mill Valley, CA: Univ. Sci. Bks., 1989.

## Collections and Anthologies

Brockman, R. John, and Fern Rood, eds. *Technical Communication and Ethics,* collection of 16 articles and papers. Arlington, VA: Soc. Tech. Com., 1989.
Girill, T. R., ed. *Among the Professions,* 26 essays reprinted from *Technical Communication* journal. Arlington, VA: Soc. Tech. Com., 1991.
Sides, Charles H., ed. *Technical and Business Communication,* collection of essays focusing on the challenges of technical communication. Arlington, VA: Soc. Tech. Com., 1989.
Sullivan, Frances J., ed. *Basic Technical Writing,* 25 papers presented at STC Annual Conferences. Arlington, VA: Soc. Tech. Com., 1987.

## Journals and Periodicals

This is a short list of periodicals which are most likely to feature articles of interest to the technical writer. Additional publications will probably be found at your public or school library. A listing of current articles should also be available. Consult a librarian for sources.

*American Medical Writers' Association Journal* (quarterly). Published by AMWA, 9650 Rockville Pike, Bethesda, MD 20814.
*College English* (monthly). Published by National Council of Teachers of English, 1111 Kenyon Road, Urbana, IL 61801.

*Editor & Publisher: The Fourth Estate* (weekly). Published by Robert Brown, Editor & Publisher Co., 11 West 19th Street, New York, NY 10011.

*IEEE Transactions on Professional Communication.* Published by Institute of Electrical & Electronics Engineers, 345 E. 47th Street, New York, NY 10017.

*Journal of Business Communication* (quarterly). Bulletin of the Association for Business Communication, published by Department of Management, College of Business Administration, University of North Texas, Denton TX 76203.

*Journal of Technical Writing & Communication* (quarterly). Published by Baywood Publishing Co., 26 Austin Avenue, P.O. Box 337, Amityville, NY 11701.

*Library Journal.* Published by Cahners Publishing Co., Bowker Magazine Group, 249 West 17th Street, New York, NY 10011.

*Publisher's Weekly.* Published by Cahners Publishing Co., 249 West 17th Street, New York, NY 10011.

*Technical Communication* (quarterly). Published by Society for Technical Communication, 901 Stuart Street, Arlington, VA 22203.

*The Technical Communication Quarterly* (formerly *The Technical Writing Teacher*). Published by the Rhetoric Department, University of Minnesota, 202 Haecker Hall, St. Paul, MN 55108.

*Technology Review* (8 issues/year). Published by the Massachusetts Institute of Technology, Building W59, MIT, Cambridge, MA 02139.

*Writer's Digest: Your Monthly Guide to Getting Published.* Published by F & W Publications, 1507 Dana Avenue, Cincinnati, OH 45207.

*Writing Instructor* (quarterly). Published by Freshman Writing Program, University of Southern California, Los Angeles, CA 90089.

# ORGANIZATIONS AND ASSOCIATIONS

Accrediting Council on
Education in Journalism & Mass
Communications
  University of Kansas
  School of Journalism
  Lawrence, KS 66045

American Advertising Federation
1101 Vermont Avenue NW
Suite 500
Washington, DC 20005

American Association of
Advertising Agencies
  666 3rd Avenue, 13th Floor
  New York, NY 10017

American Association of
Agricultural Communicators of
Tomorrow
  University of Illinois
  67 Mumford Hall
  1301 W. Gregory
  Urbana, IL 61801

American Booksellers
Association
  560 White Plains Road
  Tarrytown, NY 10591

American Medical Writers'
Association
  9650 Rockville Pike
  Bethesda, MD 20814

American Society of Journalists
and Authors, Inc.
  123 W. 43rd Street
  New York, NY 10036

Associated Business Writers of
America
  1450 S. Havana, Suite 620
  Aurora, CO 80012

Association of American
Publishers
  220 East 23rd Street
  New York, NY 10010

Association for Business
Communication
  Department of Management
  College of Business
  University of North Texas
  Denton, TX 76203

Authors Guild, Inc.
  330 W. 42nd Street
  New York, NY 10036

Aviation/Space Writers'
Association
  17 S. High Street, Suite 1200
  Columbus, OH 43215

Business/Professional
Advertising Association
  100 Metroplex Drive
  Edison, NJ 08817

Copywriter's Council of
America
  Communications Bldg. 102
  Seven Putter Lane
  Middle Island, NY 11953

Council for the Advancement of
Science Writing
  Abbotts Building, Room 100
  Philadelphia, PA 19104

Council of Biology Editors
  One Illinois Center, No. 200
  111 E. Wacker Drive
  Chicago, IL 60601

Dow Jones Newspaper Fund
  P.O. Box 300
  Princeton, NJ 08543

Editorial Freelancers Association
  P.O. Box 2050
  Madison Square Station
  New York, NY 10159

Freelance Editorial Association
  P.O. Box 835
  Cambridge, MA 02238

Health Sciences
Communications Association
  6728 Old McLean Village
    Drive
  McLean, VA 22101

Magazine Publishers Association
  575 Lexington Avenue
  New York, NY 10022

National Association of
Agricultural Journalists
  c/o Audrey Mackiewitz
  312 Valley View Drive
  Huron, OH 44839

National Association of Black
Journalists
  P.O. Box 17212
  Washington, DC 20041

National Association of
Government Communicators
  669 S. Washington Street
  Alexandria, VA 22314

National Association of Hispanic
Journalists
  National Press Building
  Washington, DC 20045

National Association of Home
and Workshop Writers
  c/o Alfred Lees
  140 Nassau Street, Room 9B
  New York, NY 10038

National Association of Science
Writers, Inc.
  Box 294
  Greenlawn, NY 11740

National Council of Teachers of
English
  1111 Kenyon Road
  Urbana, IL 61801

National Federation of Press
Women
  Box 99
  Blue Springs, MO 64013

National Writers' Club
  1450 S. Havana, Suite 620
  Aurora, CO 90012

National Writers Union
  873 Broadway, Suite 203
  New York, NY 10003

Newspaper Farm Editors of
America
  4200 12th Street
  Des Moines, IA 50313

The Newspaper Guild
  8611 2nd Avenue
  Silver Spring, MD 20910

Public Relations Society of
America
  33 Irving Place, 3rd Floor
  New York, NY 10003

Science Fiction Writers of America
Five Winding Brook Drive, No. 18
Guilderland, NY 12084

Society of Professional Journalists (Sigma Delta Chi)
16 South Jackson
Greencastle, IN 46135

Society for Technical Communication
901 Stuart Street, Suite 304
Arlington, VA 22203

Western Writers of America, Inc.
c/o Francis Fugate
2800 N. Campbell
El Paso, TX 79902

Women in Communications, Inc.
2101 Wilson Boulevard, Suite 417
Arlington, VA 22201

Writer's Alliance
Box 2014
Setauket, NY 11733

Writers Guild of America, West
8955 Beverly Boulevard
Los Angeles, CA 90048

## CANADIAN ASSOCIATIONS

Canadian Authors Association
121 Avenue Road, Suite 104
Toronto, ON, M5R 2G3
(Publishes *Canadian Author & Bookman*—quarterly)

Writers' Alliance of Newfoundland and Labrador
P.O. Box 2681
St. John's, NF, A1C 5M5

Writers' Federation of New Brunswick
P.O. Box 37, Station A
Fredericton, NB, E3B 4Y2
(Publishes *Writers News*—bi-monthly)

Writers' Federation of Nova Scotia
#203, 5516 Spring Garden Road
Halifax, NS, B3J 1G6

Writer's Guild of Alberta
10523 100 Avenue
Edmonton, AB, T5J 0A8

Writers Union of Canada
24 Ryerson Avenue
Toronto, ON, M5R 2G3

## INTERNATIONAL ASSOCIATIONS

International Federation of Scientific Editors' Associations
c/o Dr. I Elizabeth Zipf
Bioscience Information Service
2100 Arch Street
Philadelphia, PA 19103

International Society for Technology in Education
University of Oregon
1781 Agate Street
Eugene, OR 97403

International Women's Writing Guild
Box 810, Gracie Station
New York, NY 10028